Aquariums

A FIREFLY BOOK

Published by Firefly Books Ltd. 2005

First printing

Publisher Cataloging-in-Publication Data (U.S.)

Maître-Allain, Thierry.

Aquariums : the complete guide to freshwater and saltwater aquariums/ Thierry Maître-Allain and Christian Piednoir.

Originally published as: Le grand guide de l'aquarium ; France: Sélection du Reader's Digest, 2003.

[228] p. : col. photos. ; cm.

Summary: Tips for tropical freshwater fish keepers, including information on aquariums, heating and lighting, water management, aquarium plants, aquascaping, feeding, compatibility, maintenance and healthcare.

ISBN 1-55407-085-6

1. Aquariums. I. Piednoir, Christian. II Title.

639.34 dc22   SF457.M35 2005

Library and Archives Canada Cataloguing in Publication

Maître-Allain, Thierry

Aquariums : the complete guide to freshwater and saltwater aquariums/ Thierry Maître-Allain and Christian Piednoir.

Translation of: Le grand guide de l'aquarium.

Includes index.

ISBN 1-55407-085-6

1. Aquariums. 2. Marine aquariums. I. Piednoir, Christian II. Title.

SF457.M2513 2005   639.34   C2005-903409-2

Published in the United States by
Firefly Books (U.S.) Inc.
P.O. Box 1338, Ellicott Station
Buffalo, New York 14205

Published in Canada by
Firefly Books Ltd.
66 Leek Crescent
Richmond Hill, Ontario L4B 1H1

English translation by Matthew Clarke for First Edition Translations Ltd., Cambridge, U.K.

Cover design by Sideways Design

Cover photographs by Interpet Publishing

Printed in China through Printworks Int. Ltd.

Aquariums

The Complete Guide to Freshwater and Saltwater Aquariums

Thierry Maître-Allain
and
Christian Piednoir

FIREFLY BOOKS

CONTENTS

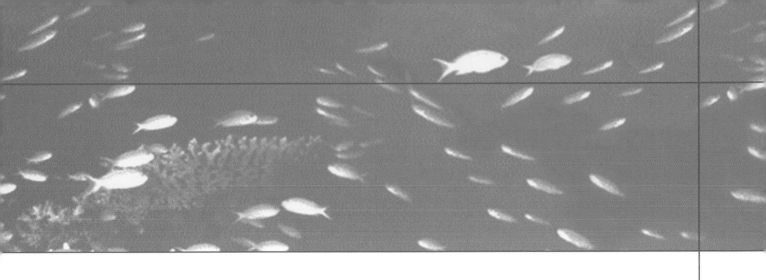

SETTING UP AN
AQUARIUM

WATER

Fish and plants are surrounded by water in the enclosed environment of a tank. If they are to flourish, an aquarist must fully understand the nature of this environment. Just as we are sensitive to the purity of the air we breathe, the characteristics of water exert a crucial influence on fish. In order to obtain water of the highest quality, it is vital to be aware of its various basic parameters so that they can be monitored. The water on the earth does not possess the same chemical qualities everywhere. In aquatic terms, there is an enormous difference, for example, between the Amazon and Central America, even though these two regions are relatively close geographically. The water in the former is soft and acidic, while that of the latter is hard and alkaline – yet fish thrive in both of them! The native species of each region have long since adapted to their respective environments, but neither could survive in an aquarium with less extreme water conditions. The capacity to control the water of an aquarium is therefore an essential prerequisite for its smooth functioning and the maintenance of healthy fish and plants.

The water cycle

Water is a liquid that changes under the influence of biological and chemical processes, which can modify its characteristics. It is important to know how to control water quality and thereby create the environment most suited to your fish.

A perpetual cycle

Water in nature is continually moving. It travels from the atmosphere, through lakes, along rivers and across oceans, before going back to its starting point.

This cycle is triggered by the sun, which causes water to evaporate from all bodies of water, particularly the ocean. Vapor is also given off by plants, in a process known as transpiration.

Evaporated water is pure and virtually without contaminants, but it does not stay that way for long. All this water condenses at high altitudes to form clouds; when these pass over land, they become denser, and the vaporized water turns to liquid – in other

H_2O

Water molecules
Molecules of pure water are formed from two hydrogen atoms linked to one oxygen atom. Chemical substances and environmental factors then endow it with specific characteristics.

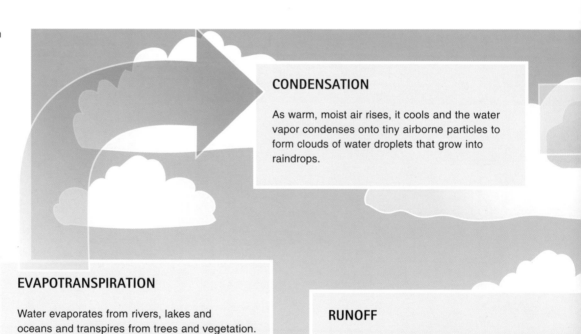

CONDENSATION

As warm, moist air rises, it cools and the water vapor condenses onto tiny airborne particles to form clouds of water droplets that grow into raindrops.

EVAPOTRANSPIRATION

Water evaporates from rivers, lakes and oceans and transpires from trees and vegetation. Only pure water passes into the air, leaving any chemicals behind.

RUNOFF

As the water reaches the ground, it flows through and over the topsoil. Humic acids and carbon dioxide make it more acidic and minerals give the water hardness.

How does water circulate?

words, to rain. The rainwater takes on oxygen and carbon dioxide (causing slight acidification), but also pollutant particles and substances.

When the rain reaches the ground, part of the water flows down slopes to join rivers, while the rest filters into the earth, where it is absorbed by plant roots. During its journey overland, the water picks up organic compounds from vegetation that render it soft and acidic, or minerals that turn it hard and alkaline. It can also, unfortunately, pick up pollutants, such as fertilizers and pesticides, from agricultural land.

After it rains, the percolation of the water depends on the characteristics of the soil. Here, once again, it can absorb minerals and become harder. It ends up forming groundwater, which may reach the surface and become a spring. Eventually, water arrives back in the oceans, and the cycle starts all over again.

For millions of years, rivers have transported organic matter and materials to the sea, thereby affecting its salinity. Fine sediment that is left behind forms silt at the river mouth. The salinity here is variable, as there is a mixture of freshwater and saltwater, and specific fauna and flora have evolved to adapt to these circumstances.

The water circulating around the earth is therefore always the same, continually performing the same cycle, but its characteristics vary along the way.

The invisible characteristics of water

RAIN

As it falls, rainwater begins to change. Pollutants and airborne chemicals are picked up on the way down, making the water slightly acidic.

PERCOLATION

Water percolating through rocks picks up nutrients and minerals. If it reaches an impermeable layer, it collects underground and emerges as a natural spring.

Nitrogenous substances
These are derived from natural chemical and biological processes, or from pollution.

Temperature
This affects oxygen levels. Most tropical fish live at 72–79°F (22–26°C).

Chlorine/ Chloramine
This is added to tap water to make it potable. It is toxic for fish but can be removed by adding a dechlorinator or by vigorous aeration.

Oxygen
This is essential to all living things. It comes from the air and diffuses into water more easily when the surface is agitated.

Carbon dioxide
Produced by the respiration of animals and plants, carbon dioxide is lost from water at the surface.

Hardness
The levels of calcium and magnesium salts render water hard or soft.

Salinity
This is a measure of the total quantity of salts dissolved in water. Sea water has about 35 grams per liter.

O₂

Oxygen
This gas is made up of two oxygen atoms and comprises 20 percent of the air we breathe. It is essential to the survival of almost all living organisms.

CO_2

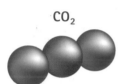

Carbon dioxide
This is made up of two oxygen atoms linked to one carbon atom. When dissolved in water, it produces carbonic acid.

Oxygen and carbon dioxide
All living organisms respire ("breathe") – not only animals but also plants. They therefore need oxygen (O_2), which is present in dissolved form in water, but in less abundance than in the air. Agitating the water surface helps oxygen dissolve, and there are various ways of doing this, such as placing the filter outlet at the surface of the water or using diffusers linked to an air pump. Keep in mind that plants give off oxygen by day, through photosynthesis (this phenomenon is described in the section on aquarium plants, pages 258–59).

Living organisms not only use oxygen to respire, they also produce carbon dioxide (CO_2). This is a toxic product, but it never accumulates in an aquarium. The process of agitating the water for oxygenation encourages carbon dioxide to pass into the atmosphere. Carbon dioxide is also absorbed in the daytime by plants through photosynthesis.

Temperature and oxygen

When the temperature goes up, the level of dissolved oxygen decreases. Summer heat can cause the temperature of an aquarium to soar above acceptable levels, and in such cases it is advisable to increase the oxygenation of the water.

The oxygen cycle in an aquarium

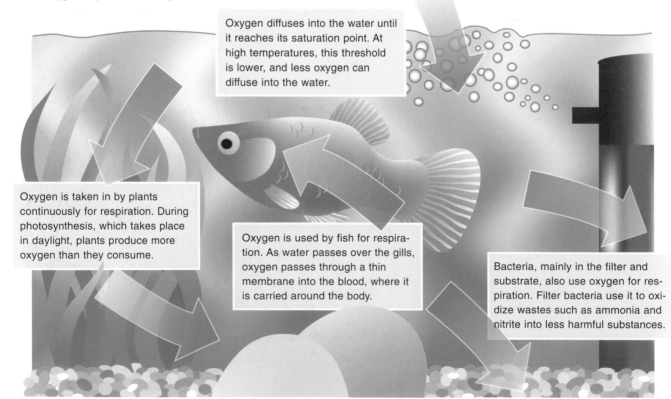

Oxygen diffuses into the water until it reaches its saturation point. At high temperatures, this threshold is lower, and less oxygen can diffuse into the water.

Oxygen is taken in by plants continuously for respiration. During photosynthesis, which takes place in daylight, plants produce more oxygen than they consume.

Oxygen is used by fish for respiration. As water passes over the gills, oxygen passes through a thin membrane into the blood, where it is carried around the body.

Bacteria, mainly in the filter and substrate, also use oxygen for respiration. Filter bacteria use it to oxidize wastes such as ammonia and nitrite into less harmful substances.

Photosynthesis

Plants have special cells called chloroplasts that use light energy to combine water and carbon dioxide molecules into glucose. This process is known as photosynthesis. Oxygen is produced as a "waste product" of this reaction.

■ Oxygen

■ Carbon dioxide

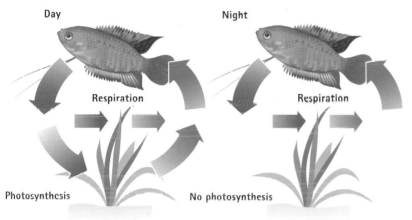

Day

Night

Respiration

Respiration

Photosynthesis

No photosynthesis

By day, plants photosynthesize, using up carbon dioxide and producing oxygen. Plants also respire, using less oxygen than produced in photosynthesis and expelling less carbon dioxide than used up in photosynthesis.

At night, plants cannot photosynthesize due to the absence of light energy, but they continue to respire. At this time, both plants and fish are using up oxygen for respiration and producing carbon dioxide.

Tap water and chlorine

The vast majority of aquariums are filled with tap water, which is treated to make it potable, often with chlorine (Cl_2) gas or its derivatives, such as chloramine. Tap water sometimes bears traces of this chlorine, to the detriment of aquatic life forms. If the water is allowed to stand for around 24 hours, the chlorine should disappear into the atmosphere (where it no longer poses a threat to the tank environment). Chloramine, however, cannot be removed by aeration alone – special products available in aquarium stores must be used. In both cases, aeration makes it easier to eliminate this undesirable element.

The carbon dioxide cycle in an aquarium

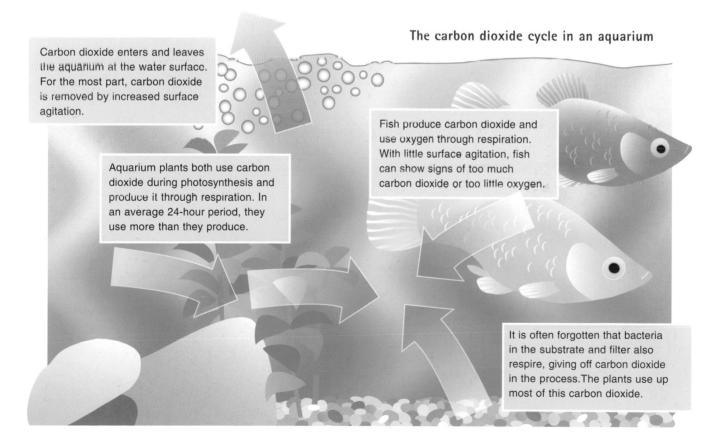

Carbon dioxide enters and leaves the aquarium at the water surface. For the most part, carbon dioxide is removed by increased surface agitation.

Aquarium plants both use carbon dioxide during photosynthesis and produce it through respiration. In an average 24-hour period, they use more than they produce.

Fish produce carbon dioxide and use oxygen through respiration. With little surface agitation, fish can show signs of too much carbon dioxide or too little oxygen.

It is often forgotten that bacteria in the substrate and filter also respire, giving off carbon dioxide in the process. The plants use up most of this carbon dioxide.

Water hardness

The hardness of water is the measurement of its levels of calcium and magnesium salts. Hard water is rich in calcium salts, and is sometimes described as being calcareous.

Soft water, such as rain, has low levels of these salts.

The total hardness is the sum of the permanent hardness, derived from certain salts, including calcium carbonate (or limestone), and the temporary hardness, mainly derived from calcium bicarbonates. This temporary hardness disappears once water has been boiled. The total hardness is measured in degrees TH or degrees GH. Hardness can be measured on the spot by means of simple color tests.

The hardness of water is linked to its pH. Soft water is often acidic, whereas hard water is generally basic, or alkaline. Tap water is rarely soft and can often be calcareous. Although it is easy to increase the hardness of water, it is more difficult to reduce it, unless a reverse-osmosis unit is used.

Hardness is easy to measure with a color test. Here, a cardboard strip soaked in water is compared to a color scale designed with the accuracy needed for aquarium testing. Other parameters are also recorded on this test strip.

A reverse-osmosis unit

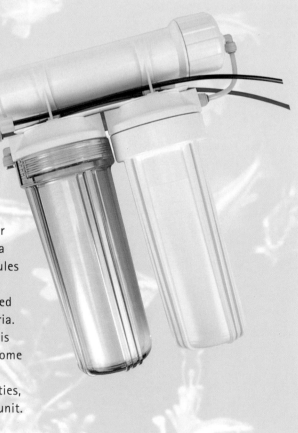

Very soft or only slightly hard water can be found in natural springs, but these can be hard to find and are often polluted. Another alternative is to collect rainwater in an inert container; however, it may contain pollutants. The simplest solution is to use water produced by a reverse-osmosis (RO) unit. This device pushes untreated water under pressure through a membrane with a mesh so fine that only water molecules can pass through, while most of the dissolved salts and metals are trapped by the membrane, along with bacteria. If only a small amount of RO water is required, it is preferable to buy it; some aquarium clubs and dealers stock it. If you are dealing with large quantities, then it is worth investing in an RO unit.

CHANGING WATER HARDNESS

Creating soft water

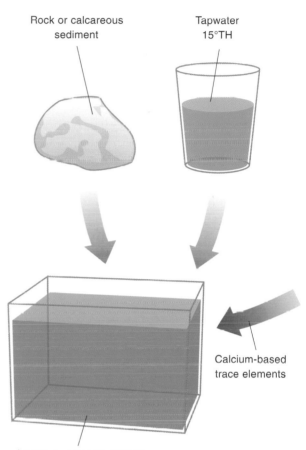

Reverse-osmosis water
0°TH

Dechlorinated tap water
15°TH

pH buffer

Aquarium water 7.5°TH

A mixture of 50 percent RO water with zero
hardness and 50 percent tap water at 15°TH
produces a fairly soft water of 7.5°TH.
The hardness can be adjusted by varying
the proportions.

Creating hard water

Rock or calcareous
sediment

Tapwater
15°TH

Calcium-based
trace elements

Aquarium water 20–30°TH

Calcareous rocks and sediment can be added
to water to increase its hardness; they must be
removed when the desired value has been
obtained. Increased hardness is paralleled by
a rise in the pH – hard water is therefore alkaline.

Creating soft
or hard water

To make soft water, all you have to
do is mix tap water with other water
of negligible hardness until you obtain
the required hardness. To create hard
water – a more unusual exercise in
an aquarium – place calcareous rocks
or oyster shells in the water, then
take them out when the hardness is
at the right level.

The pH level

The pH level indicates whether water is acidic or alkaline. In the latter case, water can also be defined as "basic."

This is a very important parameter, as some fish can only live in acidic water (South American species) or alkaline water (those from the great African lakes) and are sensitive to any major variations in the pH. However, a great many fish can adapt to water with a different pH to which they are accustomed, though they will often display more muted colors and be more susceptible to disease, and may prove difficult to breed.

The pH value is not particularly stable in aquariums, especially if plants are present. At night,

Buffering power

This is the capacity of water to maintain its pH at a particular value, without any significant fluctuations. Buffering power is largely derived from the presence of limestone. This is why hard water and seawater have such great buffering power and display so few variations in their pH. In contrast, soft, acidic water is subject to more substantial variations in its pH. Buffering power can be created in an aquarium by placing calcareous rocks on the bed or in the decor.

Acidifying water

This is often vital for some fish, particularly when trying to breed them. The buffering power of hard water makes acidification difficult, but it is much easier in soft or reverse-osmosis water, which is simply filtered over peat until the desired pH level is obtained.

plants produce carbon dioxide, which acidifies the water and causes the pH to drop, but by day they consume this gas and the pH rises. Such variations are entirely normal.

The pH can, however, plummet in the long term, over the course of a few months. This is due to waste residues emanating from the fish and plants. Good filtration and periodic water changes help to prevent this problem – as long as the pH of any added water is the same as that of the aquarium. It is very easy to measure the pH level with a simple color test, which should be carried out on a fairly regular basis.

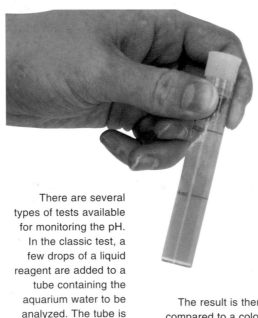

There are several types of tests available for monitoring the pH. In the classic test, a few drops of a liquid reagent are added to a tube containing the aquarium water to be analyzed. The tube is then shaken for a few moments.

The result is then compared to a color scale provided with the test. In this case, the measurement shows that the water is very slightly acidic.

5.5

6.1

6.4

6.7

7.0

7.5

HOW pH CHANGES OVER A 24-HOUR PERIOD

During the day
Carbon dioxide acidifies the water. The plants absorb this for photosynthesis, so the water is less acidic: the pH goes up but stays within normal values and has no effect on the fish.

At night
There is no photosynthesis; the plants and fish respire and give off carbon dioxide, thereby acidifying the water and lowering the pH value. All these variations are normal.

The pH measured at the end of the day represents a maximum value.

In contrast, the pH is at its lowest level at the end of the night.

The pH scale

pH 9: the water is 100 times more alkaline than at pH 7

pH 8: the water is 10 times more alkaline than at pH 7

pH 7: neutral

The pH scale is logarithmic. Each change in the pH number corresponds to a factor of 10: water at pH 8 is 10 times more alkaline than water at pH 7; water at pH 9 is 100 times more alkaline than water at pH 7. This is why a sudden change in the pH level is very stressful and harmful to aquarium fish.

The pH scale

The pH scale ranges from 0 to 14 (there are no units). With very few exceptions, fish can only survive within the range of 5 to 9 in a natural setting. Neutral water has a pH of 7. Below that, water is acidic; higher up, it is alkaline. It is unusual for the pH of a freshwater aquarium to drop below 6 or rise above 8.

Seawater is alkaline, however, due to the salts it contains, and its pH must not stray far from 8.3.

The pH is measured by colored indicators that are easy to use. In many test kits, yellow signifies acidic water, green, neutral water, and blue, alkaline water.

Temperature

Most tropical fish should be raised at a temperature of 72°F to 79°F (22°C to 26°C), although some species can tolerate slightly lower or higher temperatures. It is easy to warm up aquarium water with an electric heater; the models currently on the market are accurate to about 2°F (1°C). A slight variation between daytime and nighttime temperatures is normal, but any sharp drop can be harmful to fish and may trigger disease. When the water is changed, it is therefore important to ensure that the temperature of the new water is the same – more or less – as that of the aquarium. Sometimes, however, an increase of a few degrees can be beneficial to fish, particularly when they are being encouraged to reproduce.

Digital thermometers containing liquid crystals can be attached to the exterior of the aquarium (their rear surface is adhesive). They make it very easy to read the temperature.

Glass models containing alcohol are placed inside the aquarium. They are attached with a suction cup away from the heater, to avoid being directly influenced by it.

The effect of the temperature on the toxicity of ammonia

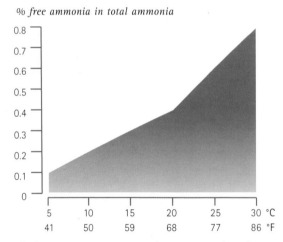

% free ammonia in total ammonia

Ammonia occurs in two forms: free ammonia (NH_3) and ammonium (NH_4^+). Free ammonia is the most toxic, but it only accounts for a tiny part of total ammonia. Their proportions vary in relation to the temperature: the warmer it is, the higher the levels of free ammonia. The pH also exerts an influence on the free ammonia content, as this increases parallel to any rise in the pH. However, free ammonia rarely exceeds a level of 1 percent with the temperatures and pH values normally found in an aquarium.

The influence of the temperature on the oxygen levels

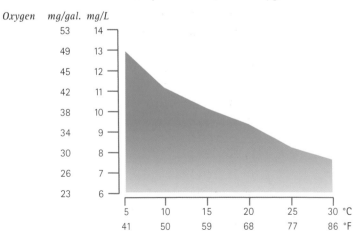

Oxygen   mg/gal.   mg/L

When the water temperature rises, there is a drop in the amount of oxygen that can be dissolved. The oxygen level varies by about 20 percent between temperatures of 59°F and 77°F (15°C and 25°C). At the same time, the metabolism of fish, plants and bacteria speeds up, increasing their need for oxygen. Therefore, there is a risk of an oxygen deficit at high temperatures. This is why it is vital to avoid overheating and overpopulating an aquarium. If oxygen levels drop, do not hesitate to boost the oxygenation of the water by vigorously agitating it.

The density of seawater

The saltier seawater is, the greater its density (specific gravity), but this decreases as temperature rises. Freshwater is the constant (1.000) against which the density of other liquids is measured, with seawater measuring 1.025. In saltwater aquariums, where the temperature is relatively stable, specific gravity will only be altered by the degree of salinity. Specific gravity can be measured using a hydrometer, traditionally a small device that floats upright and becomes increasingly buoyant as the water becomes saltier. The hydrometer scale can then be read at the surface, a method sufficiently precise for aquarists. Some hydrometers also contain a thermometer, since temperature is a vital part of accurately interpreting specific gravity. Floating-arm hydrometers are widely used.

Seawater

Seawater is comprised of a number of salts dissolved in certain proportions; in all, it contains over 50 chemical elements.

The main salt is sodium chloride (the same salt used for cooking). However, seawater cannot be obtained by dissolving salt in tap water.

Whatever the sea, the proportion of salts is always the same – it is only the total quantity that varies. Salt content is generally about 34.5 ppt (parts per thousand) in seawater. This figure is somewhat lower nearer coastlines, due to the influx of freshwater, and considerably reduced near the poles. In tropical areas – the source of most saltwater aquarium fish – it can rise to 35–36 ppt (sometimes slightly more in the Red Sea). It is easy to reconstitute seawater in an aquarium by using the special salts available on the market, always following the dosage specified in the manufacturer's instructions. You can also check whether the salinity in a tank is appropriate by measuring the specific gravity of the water. At a temperature of 75–79°F (24–26°C), the density should be 1.023–1.024 in a saltwater aquarium.

Nitrogenous substances

Nitrogenous substances are the byproducts of biological processes. Some are very dangerous for animals, others less so. They all contain nitrogen (N), a major element in living matter that is found particularly in the proteins that make up muscle. Nitrogenous substances can be measured with colored indicators that are extremely easy to use. It is advisable to monitor them on a regular basis.

The ammonia level must be checked regularly, especially in a tank that has recently been set up. Aquarium stores sell color tests, incorporating either a liquid reagent or tablets that are dissolved in a sample of water.

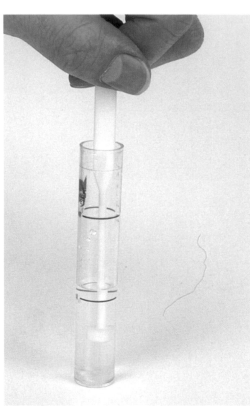

Ammonia

Ammonia (NH_3) is derived from the excretions of fish or the decomposition of organic matter (surplus food or plant debris, for example). It is highly toxic: a concentration of 0.2 milligrams per liter (mg/liter) kills fish immediately, but even a level 10 times lower has the same effect over the long term. Ideally, therefore, there should be no ammonia in a tank at all. Free ammonia (NH_3) is more toxic than ammonium (NH_4^+) but is only found in small quantities.

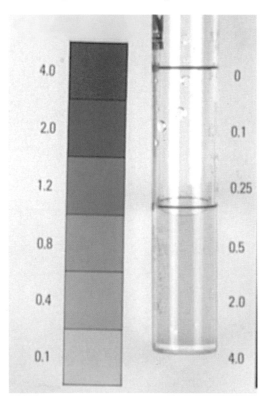

The color obtained in a water sample is compared to a scale to determine the level of ammonia. In this case, it is close to 0.1 mg/liter (about 0.4 mg/gal.).

NH_3

Hydrogen (H)

Nitrogen (N)

Ammonia
The free ammonia molecule consists of three hydrogen atoms and one nitrogen atom; the ammonium molecule (NH_4^+) has an additional hydrogen atom. During the nitrogen cycle, bacteria transform ammonia into nitrites and then into nitrates by means of chemical reactions that replace hydrogen atoms with oxygen atoms.

Nitrites

Ammonia is transformed into nitrites (NO_2^-) by bacteria from the *Nitrosomonas* genus: this is the start of the nitrogen cycle. Nitrites are less dangerous than ammonia, but they are toxic nonetheless and can kill fish when they reach levels of 0.2 mg/liter (0.75 mg/gal.). Ideally, they should be totally absent from an aquarium. If the nitrite level is too high, changing the water in the tank will enable you to bring it back to a suitable level.

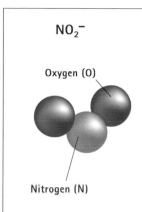

NO_2^-

Oxygen (O)

Nitrogen (N)

Nitrites

The nitrite molecule is made up of two oxygen atoms and one nitrogen atom. It is highly toxic for fish and must not be allowed to accumulate in an aquarium. Nitrites are normally transformed into nitrates by the addition of an oxygen atom. This task is performed by bacteria in the nitrogen cycle.

In this nitrite test, a tablet is dissolved in the water under analysis. In other models, a liquid reagent is used. In every case, it is essential to follow the instructions for use to ensure a correct result.

The effect of nitrites

Nitrites transform the hemoglobin in the blood and prevent it from transporting oxygen. As a result, fish are suffocated, even if the oxygen level in the water is suitable. The fine gill filaments turn from bright red to brown.

Surplus ammonia

Overly high levels of ammonia can be caused by overcrowding, overfeeding, the presence of dead fish or a defect (or breakdown) in the filtration system. The fish's gill filaments collapse; as a result, they find it hard to breathe and will take frequent gulps of air at the surface of the water. They sometimes wobble as they swim, and they will eventually die. Obviously, such circumstances require a speedy reaction. About 20–30 percent of the water must be changed daily for several days, taking care to use water identical to that in the aquarium. You must also check the filter, as it may be necessary to wash or change the filter media.

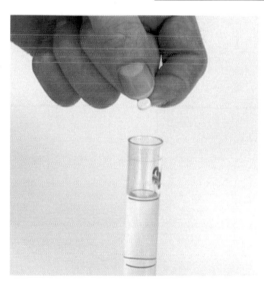

It is particularly important to measure the nitrites in a tank that has recently been set up, before the fish are introduced. Here, the result of the test shows that nitrites are totally absent, so there is no risk to the fish.

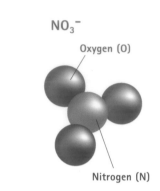

NO_3^-

Oxygen (O)

Nitrogen (N)

Nitrates

The nitrate molecule consists of three oxygen atoms joined to a nitrogen atom. It is the end product of the nitrogen cycle. Although this compound is less dangerous than ammonia or nitrites, it is nevertheless advisable to prevent it from accumulating in an aquarium. Nitrates constitute one of the mineral salts essential to the growth of plants, and they are found in fertilizers for aquarium plants.

Nitrates

Nitrates (NO_3^-) are derived from the transformation of nitrites by a second group of bacteria from the *Nitrobacter* genus. In a natural setting, runoff from rain can carry nitrates from farmland into a waterway. Nitrates are considerably less dangerous than the other nitrogenous substances mentioned above – they are only toxic at levels over 50 mg/liter (189 mg/gal.). Invertebrates, however, are markedly more sensitive to them than fish, particularly in a saltwater tank.

Nitrates mark the end of the nitrogen cycle. They are absorbed by plants, but they can accumulate over time in an aquarium. This accumulation can be reduced by changing a proportion of the water.

The nitrogen cycle

The nitrogen cycle is a biological process affecting both plants and fish that involves ammonia, nitrites and nitrates.

In the wild (with no pollution or human interference), there are generally no imbalances. In the enclosed setting of an aquarium, however, this cycle must be "powered" by means of biological filtration. In a newly set up tank, it is vital to wait until the nitrogen cycle is completely established before adding the majority of the fish. To get the cycle going, start off with two or three hardy fish in a small tank, four or five in a larger setup.

Later on, the cycle can be disrupted, particularly if there is any overcrowding or overfeeding. The nitrogen cycle depends on the activity of several types of bacteria that need oxygen to respire and thereby oxidize the ammonia and nitrite molecules. Their presence is indispensable in an aquarium.

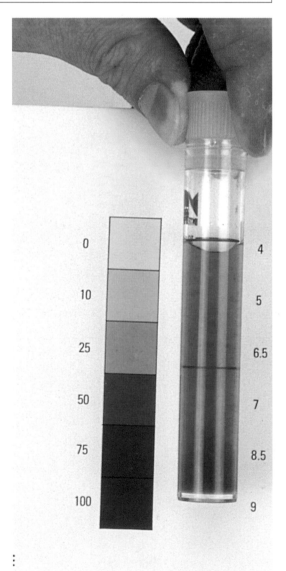

The result of this analysis reveals an excessive quantity of nitrates. Even though they are less toxic than nitrites, they must be eliminated by partial (but regular) water changes. The nitrate level can also be lowered by putting extra plants into the aquarium, as they use nitrates as food.

THE NITROGEN CYCLE

In a well-balanced aquarium that has been running for some time, the nitrogen cycle keeps the concentrations of ammonia, nitrites and nitrates at a low level. This is no excuse for an aquarist to stop making partial water changes at regular intervals, as this makes it possible to eliminate other dangerous substances that do not contribute to the nitrogen cycle. And the addition of new water introduces mineral elements that are essential for plants and fish.

Excretion
Fish do not use all the proteins from their diet, and the elimination of this surplus produces ammonia.

Food
The food eaten by fish plays a part in the nitrogen cycle, whether digested or not.

Ammonia
This is produced by the decomposition of organic material, such as debris from plants, surplus food and dead fish. Ammonia is very toxic for animals, so it must either be converted or removed.

NH_3

The influence of bacteria
Bacteria from the *Nitrosomonas* genus are the first to act. Oxygen enables them to convert ammonia (NH_3) into nitrites (NO_2^-). This is the first stage of biological filtration.

Nitrates: Fertilizers for plants
In a well-planted aquarium, plants participate in the elimination of nitrates; on occasion, they are even eaten by the fish, allowing the nitrogen cycle to come full circle.

Nitrites
Although not as toxic as ammonia, they are nonetheless dangerous even at low doses, as they reduce the blood's oxygen-carrying capacity. They are converted to nitrates as a result of biological filtration.

NO_2^-

Water changes
Regular partial water changes reduce the concentration of nitrates, and can have the same effect on ammonia and nitrites.

Nitrates
These are the end products of the nitrogen cycle. They are used by plants as nutrients.

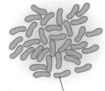

Denitrification
In areas with a limited supply of oxygen, such as compact substrate, anaerobic bacteria (which do not need oxygen) convert nitrates into nontoxic nitrogen gas (N_2).

NO_3^-

The influence of bacteria
Nitrobacter bacteria oxidize nitrites into nitrates, which are considerably less toxic. This is the final stage of biological filtration.

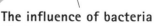

Filtration

All biological activity produces waste. In the wild, waste is highly diluted and breaks down naturally, as long as there is no pollution. In an aquarium, waste products are considerably more concentrated, so they need to be either transformed or eliminated. This is where filtration comes into play, but to be successful it must be mechanical, biological and chemical all at the same time.

A well-filtered aquarium guarantees the good health of both plants and fish.

To do this, the bacteria must be provided with a large surface area on which to grow and form colonies. Various kinds of filtration media are used, although foam is the most common. The water entering a biological filter must be well oxygenated to enhance the action of the bacteria.

Encouraging the nitrogen cycle

When an aquarium is first set up, the nitrogen cycle establishes itself on its own. This takes some time – generally four weeks in freshwater.

In fact, good bacteria are only present in small quantities (they often enter along with the plants). They have to grow if they are to colonize the filter and other surfaces.

The nitrogen cycle can therefore be encouraged by introducing bacteria into the aquarium. There are several ways to do this.

You can use products on the market that are designed to seed the filter with bacteria, or you can insert a little substrate taken from another aquarium that is already running and well balanced into your tank. This substrate will contain bacteria that will go on to colonize their new environment. You can also take advantage of a balanced aquarium by directly transferring part of its filter media – already extensively colonized by bacteria – and placing it in the filter of the tank you are setting up.

In any case, ensure that the levels of ammonia and nitrites are at zero before introducing fish into the aquarium. To do this, take regular readings of these two parameters, about twice a week.

The principle of biological filtration

Biological filtration (also known as bacterial filtration) makes it possible to recreate the nitrogen cycle and eliminate ammonia and nitrites by enhancing the work of bacteria.

Beware of power failures!

A short power failure is not dangerous, although problems can arise if it lasts longer than a few hours or if the filter breaks down (a rarer occurrence). When a biological filter is no longer supplied with water, the bacteria will die within a few hours. The filter media must therefore be washed and seeded with bacteria before the filter is put back into operation.

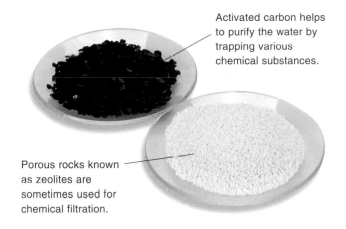

Activated carbon helps to purify the water by trapping various chemical substances.

Porous rocks known as zeolites are sometimes used for chemical filtration.

Filter wool, an artificial material with fine, intertwined filaments, traps even the smallest particles suspended in the water.

Large-grade foam will remove the larger visible particles in the aquarium. These will then be broken down by bacteria into smaller particles.

Sintered glass pieces provide an enormous surface area for beneficial bacteria to colonize.

Ceramic nodules serve the same purpose; they can substitute or complement the sintered glass pieces.

Similarly, these hollow ceramic cylinders can trap larger particles.

This is also true of these solid cylinders, which serve as a barrier to plant debris and leftover food. They will also harbor beneficial bacteria.

Chemical filtration

Water can be filtered chemically to eliminate certain substances that cannot be removed by a biological filter. The biological filter can be supplemented with active carbon, which traps any coloring emitted by wooden decor, as well as some of the atmospheric pollutants. The carbon must be changed regularly (not washed, like other filtering media).

Transparency and mechanical filtration

It is vital to make aquarium water as transparent as possible. First of all, it looks good; second, clear water enables light to penetrate right down to the lower leaves of plants. This is where mechanical filtration comes into play: various materials trap particles in suspension according to their granule size. In fact, most of these materials serve a double purpose of not only mechanical but also biological filtration, as they allow bacteria to grow on their surface and colonize rapidly.

This small, air-powered filter contains enough media to filter a small tank suitable for fry.

Bacterial starters

This liquid starter contains bacterial cultures to "kick-start" biological filters.

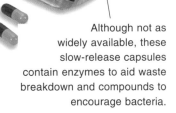

Although not as widely available, these slow-release capsules contain enzymes to aid waste breakdown and compounds to encourage bacteria.

Good bacteria

The word "bacteria" often evokes thoughts of disease. In the wild, however, some can prove very useful; these are termed "good bacteria," and they include the bacteria involved in the nitrogen cycle. In an aquarium, they can colonize various surfaces: the filter media, the substrate, the decor and even, sometimes, the plants.

27

The natural habitats of aquarium fish

In the wild, aquatic settings are extremely diverse, from slow- and fast-running rivers to stagnant ponds, from great lakes to brackish swamps, as well as the famous coral reefs. The fish that live in each setting have adapted to the local conditions, particularly the water quality. To successfully raise fish in an aquarium, it is important to recreate their original environment as closely as possible. Here are some examples of aquatic environments found around the world.

Above:
In the Amazon, the plant cover restricts light penetration, but numerous fish species thrive in these conditions. The submerged remains of plants decompose slowly and acidify the water.

The great African lakes

These great lakes are extremely deep (thousands of feet), and this large volume ensures that the characteristics of the water remain constant. The water is very hard and alkaline, but also very clear. The banks and bed consist mainly of rocks, often large blocks that are not suited to plant life. Nevertheless, some algae do grow on these rocks and contribute to the diet of herbivorous fish, which in turn become the prey of carnivorous species. This well-established food chain enables several hundred species to subsist. Some have only been discovered in the last few years, and there are certainly several more that remain undetected. Most belong to the Cichlidae family. Their temperament and behavior (some species protect their eggs and fry in their mouth) make them popular with aquarists.

The warm waters of Asia

The water in the rivers and marshes of Asia is often shallow, warm, soft, and acidic or neutral. This shallowness allows bright light to penetrate to the riverbed and

fosters lush plant life (either totally submerged or floating) that produces oxygen, which is often scarce in these settings. This is why the fish that live here require little oxygen. Some species are even equipped with an organ that complements the gills by trapping atmospheric oxygen. The most

well-known of these is the betta, or Siamese fighting fish (a name that aptly evokes the behavior of one male in the presence of another). These settings are also home to numerous species from the Cyprinidae family, cousins of the European carp. The most popular fish from this type of water is the danio (*Brachydanio* spp.), which has been the first acquisition of many an aquarist.

The acidic waters of the Amazon

Here, branches and roots clutter up the rivers and swamps while abundant rain drains the soil, resulting in the coloration and acidification of the extremely soft water (although this remains fairly transparent and permits the growth of plants). The latter are relatively scarce, however, on account of the very substantial tree cover and, therefore, limited sunlight. These waters are inhabited by small species that are often brightly colored, such as the neon tetra (*Paracheirodon innesi*) or other fish from the Characidae family. This is also the domain of more majestic species, such as the angelfish (*Pterophyllum* spp.) and discus (*Symphysodon aequifasciatus*), as well as numerous catfish and the infamous piranha, which is in fact docile when raised in an aquarium.

Above:
The bank of an African lake. The mass of rocks is partially covered by the water and so provides countless hiding places for fish, most of which are only found in the great lakes of eastern Africa.

Left:
In some parts of Asia, the marshy waters are shallow and warm. Even though oxygen is less abundant than in running water, fish have adapted to these swampy conditions.

Above:
In Central America, the water near the coast is hard. This is the realm of livebearers.

Right:
Aquatic plants are more abundant in slow-running waters, where conditions encourage dense plant growth.

The hard waters of Central America

In this rocky region, the fast rivers erode the limestone substrates, making the water hard and raising its pH. In the upper part of the river, the rapid currents and rocky bed hinder the growth of vegetation. This is the realm of certain Cichlidae, although they differ from those of the African lakes, which prefer highly oxygenated water.

In the lower depths, where the water flows more slowly, the fine sediment and clear water favor the presence of plants. The water becomes hard and alkaline, an ideal setting for the guppies and other livebearers often kept in aquariums. These fish are much appreciated by enthusiasts for their unusual form of reproduction: they do not lay eggs but give birth to fully formed fry.

Brackish waters

Here, rivers meet the sea, in estuaries and deltas, and the dividing line between freshwater and saltwater becomes blurred. The water therefore contains a certain amount of salt, according to the tides and the proximity of the sea.

There are few plants, the soil is silty and the "decor" is predominantly made up of the branches and roots of mangroves – the trees that typify this environment.

These distinctive conditions have led to highly specific wildlife that are adapted to this environment. Some freshwater species do, however, occasionally venture downstream into saltier waters.

The coral reefs

The water here is very clear and well lit, with plenty of movement and, therefore, oxygen. It is also very stable, with few chemical variations. The limited plant life is mainly confined to filamentous and macro-algae. In contrast, the fauna is very rich, as both living and dead coral reefs provide shelter for a vast number of invertebrates, such as crustaceans, worms, sea urchins and starfish.

There is a wealth of fish, most of them veritable marvels with extremely bright colors. Most saltwater aquarium fish come from these regions. They are often territorial and many cannot bear the presence of other members of their own species.

Of all these fish, the most well known is undoubtedly the clownfish, which often lives in the company of an anemone.

Above:
The clear waters and strong currents around the islands in the Pacific and Indian Oceans boast an astonishing diversity of animal life.

Below:
Corals provide many habitats that harbor countless species of fish and crustaceans.

N owadays, setting up an aquarium and making it work no longer demands any specialized knowledge. Manufacturers excel at creating ingenious ways to make an aquarist's life easier and minimize the risks of failure. Nevertheless, the first steps demand patience, whether you buy a fully set up aquarium or take on the task of building and setting up the tank yourself. It is important not to rush the initial preparations – there are no shortcuts allowed in the time between the tank positioning and the arrival of the fish! This phase involves a series of stages that all deserve meticulous consideration, but can also provide great fun: finding the best site in the house, installing the equipment and accessories, then decorating with substrate, rocks and wood before finally filling the tank with water. The aquarium must then reach maturity before it is possible to introduce large fish. This waiting period, which is connected with the nitrogen cycle, is absolutely indispensable. Unfortunately, attempts to skip this maturation period of around three weeks often lead to failure, as the fish can be exposed to high doses of ammonia and nitrites before the beneficial bacteria in the filter have increased sufficiently to process these natural but deadly products. Impatience and urgency are two words that are utterly incompatible with the set up of an aquarium!

Choosing an aquarium

Selecting a tank is not as easy as it seems, even for those with some expertise. In this chapter, we offer some advice and information, but it is advisable to visit several fish supply stores to obtain an idea of the range of models available – and their price. Ideally, seek the help of an aquarium hobbyist when choosing an aquarium.

What shape, what size?

Choosing an aquarium is, above all, a matter of taste. It requires taking into consideration not only the placement and type of support to be used, but also the quantity of fish to be kept and, obviously, the budget available.

There is a huge range of models on the market. The classic tank is rectangular in shape and is made from sturdy sheets of glass stuck together with silicone sealant to make it waterproof. The dimensions are generally standardized, but it is also possible to order a model made to your own specifications. Tanks in more unusual shapes can also be found: aquariums with a convex glass front or the corners cut off, or made to fit into the corner of a room.

Many aquariums are sold with a housing and a support, which can range from a simple metal stand (either pre-assembled or in the form of a do-it-yourself kit) to a piece of furniture with shelves and cupboards – a practical way of hiding an external filter and other equipment.

Many aquarists, however, prefer to buy a completely unadorned tank in order to integrate it into the existing style prevalent in their home.

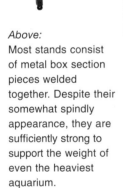

Above:
Most stands consist of metal box section pieces welded together. Despite their somewhat spindly appearance, they are sufficiently strong to support the weight of even the heaviest aquarium.

Right:
An aquarium must never be larger than its support. Most tanks are sold with appropriately sized supports in matching colors.

An aquarium is heavier than you think!

This becomes apparent when it is time to transport an empty tank, but you also need to add in the weight of the water, along with the substrate and decor, especially if either is made up of rocks. The approximate total weight (in pounds) of an aquarium filled with water only can be calculated by multiplying its volume in gallons by 8.25 (to convert into kilograms simply multiply the weight in pounds by 0.45). So, based on this weight of $8\frac{1}{4}$ pounds per gallon of water, a tank with a volume of 45 gallons will weigh over 370 pounds (165 kg)! It is therefore wise to provide a strong support. In the case of very large tanks, the floor must be strong enough to bear such a heavy load. If there is any doubt, it is well worth checking.

How much does my tank weigh?

The bigger an aquarium, the easier it will be to maintain the biological equilibrium. Moreover, the number and size of the fish must be proportional to the volume of the tank. In freshwater, a tank 32–36 inches (80–90 cm) long, with a volume of 20–30 gallons (75–115 L), is ideal for beginners. In the case of saltwater, it is not a good idea to go with a tank smaller than 65 gallons (250 L).

Dimensions of the tank (length x width x depth)	Tank volume	Total approximate weight of a full tank
24 x 12 x 12 in. (60 x 30 x 30 cm)	14 gal. (54 L)	115 lb. (50 kg)
32 x 12 x 12 in. (80 x 30 x 30 cm)	20 gal. (72 L)	165 lb. (75 kg)
36 x 14 x 14 in. (90 x 35 x 35 cm)	30 gal. (110 L)	250 lb. (115 kg)
40 x 16 x 16 in. (100 x 40 x 40 cm)	45 gal. (170 L)	370 lb. (165 kg)
48 x 16 x 20 in. (120 x 40 x 50 cm)	63 gal. (240 L)	520 lb. (235 kg)
60 x 20 x 20 in. (150 x 50 x 50 cm)	100 gal. (375 L)	825 lb. (375 kg)
80 x 20 x 20 in. (200 x 50 x 50 cm)	132 gal. (500 L)	1,090 lb. (495 kg)

Whatever tank is chosen, however, it should never pass beyond the edges of its support. As well, it is always advisable to mask the upper part of the tank – this hides the surface of the water and makes the aquarium more visually attractive.

Some models come complete with filtration and heating equipment and various accessories. This may seem ideal, but it is sometimes wise to pass on such offers as they may involve taking on items that are either too small or too big.

Below:
Aquariums set into a custom-made piece of furniture are extremely practical, as equipment can be hidden underneath the tank. External claddings are available in a variety of styles, thereby allowing an aquarium to blend into its surroundings.

Bargain offers

Secondhand aquariums can be found through classified ads in local newspapers, where they are often offered complete with their equipment. These aquariums are a tempting offer but they must be approached with caution. Before buying, you must examine the state of the glass (no scratches) and the glued joints, and ensure that the equipment works properly. None of this is easy for a beginner to assess, so it is wise to go with a more knowledgeable enthusiast. Aquarium clubs are also a source of good bargains – along with plenty of expert advice.

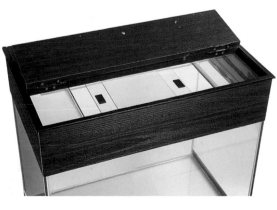

Positioning the aquarium

Once an aquarium has been set up and filled, it is not possible to move it somewhere else. It is therefore vital to consider its location carefully, as this will be more or less permanent. There are certain rules to respect, but after that it is a matter of the setting and personal taste – although the ideal position is not always easy to find.

Finding a good position for an aquarium inside a home

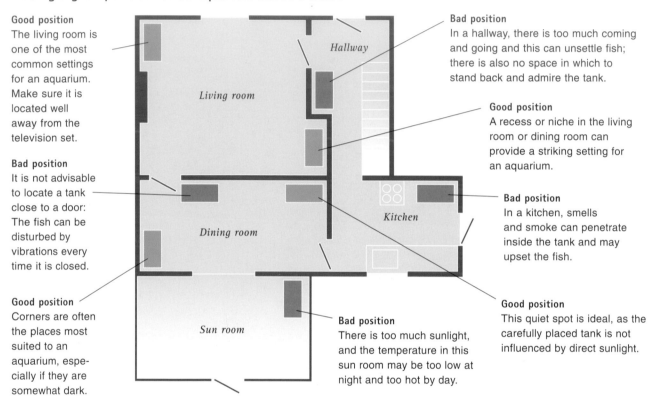

Good position
The living room is one of the most common settings for an aquarium. Make sure it is located well away from the television set.

Bad position
It is not advisable to locate a tank close to a door: The fish can be disturbed by vibrations every time it is closed.

Good position
Corners are often the places most suited to an aquarium, especially if they are somewhat dark.

Bad position
In a hallway, there is too much coming and going and this can unsettle fish; there is also no space in which to stand back and admire the tank.

Good position
A recess or niche in the living room or dining room can provide a striking setting for an aquarium.

Bad position
In a kitchen, smells and smoke can penetrate inside the tank and may upset the fish.

Good position
This quiet spot is ideal, as the carefully placed tank is not influenced by direct sunlight.

Bad position
There is too much sunlight, and the temperature in this sun room may be too low at night and too hot by day.

Hallway

Living room

Kitchen

Dining room

Sun room

While the tank is still empty, it is well worth taking time to move it around with its support and try out several different locations. It is very important to bear in mind the placement of the tank in relation to the floor. Ideally, an aquarium should be at an adult's eye-level when sitting down; this also allows young children to observe the fish. Furthermore, it is easier to maintain a tank of this height than one set higher up.

Aquariums are frequently installed in a living room or dining room, particularly in corners, recesses or alcoves. As these are the darkest parts of a room, they tend to set off an aquarium superbly.

Places to avoid

Hallways or areas subject to a great deal of coming and going are not suited to aquariums. Heavy footsteps and banging doors create vibrations; fish can detect these and be disturbed by them. This is especially true when they are breeding.

Possible sites

Children making their first venture into the world of aquariums often want to keep the tank in their bedroom. There is no reason why not, apart from the slight noise of the equipment at night.

A basement is another possible alternative, provided it is not too hot in summer, as an aquarium is difficult to cool down. It is easier to heat it up if the rooms are too cold in winter: in this case, provide a level of heating that exceeds the normal recommendations based on the capacity of the tank.

A garage can also be a suitable site, but with the same considerations as above. Some aquarists use their garage to keep a set of quarantine and breeding aquariums.

A well-stabilized aquarium

Many aquariums come with a plastic base, which will absorb any irregularities in the support, and thereby prevent the tank from tilting and the glass pane that forms the bottom from cracking under stress. If your aquarium does not have a plastic base do not place it directly on top of the support or cabinet. First position a flat board

between the tank and the support, topped with a thick sheet of Styrofoam or a plastic foam mat.

Finally, use a level to check that the whole setup is completely horizontal. Reposition as necessary – some supports on the market have adjustable feet to make this easier to do.

It is vital to place a plastic base or a Styrofoam sheet about half an inch thick between the support and the bottom of the aquarium, to compensate for any possible irregularities in the support.

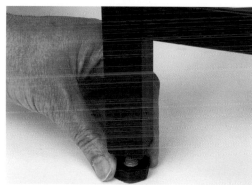

Left:
Use the adjustable feet found on some stands to set the tank completely level.

Below:
The aquarium can be checked with a level to ensure it is level (do this before filling the tank).

Some rules to follow

In order to eliminate the need for extension cords, the aquarium should be fairly close to a power outlet. It should also be readily accessible, not only for the installation but also for maintenance; it is particularly important to ensure that the hood can be removed easily. Do not position an aquarium close to a heat source, such as a radiator or fireplace, and also avoid putting the tank right in front of a window, particularly if it faces south. This can raise the temperature of the water, as well as encourage the growth of undesirable algae. Furthermore, direct sunlight produces an effect that is visually unattractive. It is also inadvisable to place the tank too close to a television. Not only will the fish be constantly disturbed by the sound vibrations, but also you will not know which one to watch!

The aquarium in the home

1 Some aquariums are designed to be placed in a corner. The position of this 25-gallon (95 L) tank between the sofa and the armchair creates a restful atmosphere.

2 This 65-gallon (250 L) tank is typical of those sold complete in pet stores. It rests on a black, melamine-coated wooden support specially made to bear a weight of at least 660 pounds. (300 kg).

3 When interior decoration is considered, the resulting aquarium can be as elegant as this blue and white ensemble. This 6$^1/_2$-foot (2 m) saltwater tank was custom-made to fit between the two windows.

4 A built-in aquarium is undoubtedly the most attractive option of all. Unfortunately, it is not always possible in practice, as it is essential to have access to the back and top of the aquarium from an adjoining room.

5 This saltwater aquarium is designed to be set in the middle of a room and therefore be visible from all four sides. This original shape can only be achieved through the use of acrylic materials (which are more difficult to maintain than glass).

6 This aquarium is lit by three powerful light sources. It has been left uncovered to allow the plants to spill out, creating a striking visual effect.

1

3

6

The backdrop

Before setting up the aquarium in its final position, you can decide on what background you will attach to the outside of the back glass. The background serves to highlight the decor, plants and fish, as well as create an impression of depth. Suitable backgrounds are available in a variety of colors and photographic images.

Above:
Specialized stores offer photo backgrounds with a range of images – some of which have very little to do with the habitats of fish! The most popular ones show rocks, wood, plants or aquarium decor.

The backdrop must be attached while you still have access to the back of the tank. The sight of a wall behind the aquarium – even with the prettiest of wallpapers – is an unappealing prospect, so pet and aquarium stores sell photographic backgrounds printed on plastic sheets or rolls, showing a decor of plants, rocks, tree roots – even Greek temples! You can also buy backgrounds in plain black or graduated shades of blue (place the paler part at the top). Cut the sheet or roll to size and attach it to the outside of the back glass with tape. A black background is a safe choice – it gives the finished aquarium a sense of depth and sets off the plants and fish.

Right:
Attach the background outside the tank's back glass.

Above:
This plastic background comes in a roll and can be cut to fit the rear glass panel exactly. This example offers a choice of graduated blue or plain black.

The freshwater substrate

The choice of substrate for the freshwater aquarium is a matter of taste, but it also depends on the fish, plants and type of filter being used in the aquarium.

Substrate texture

Sand and gravel of various particle sizes (grades) are available in specialized stores, but the most suitable substrate is sand with round, eroded grains. Some fish rummage in the substrate for food, and it is best to provide these with fine or quartz sand. A coarser substrate is required, however, if you are using an undergravel filter (see pages 46–47).

Adding the substrate

The substrate must be rinsed until the water is clear, then carefully spread over the bottom of the aquarium. If you are using an undergravel filter, the filter plates must be placed on the base of the tank before the substrate. A slight gradient from the front toward the back makes the tank appear deeper. The average depth of the substrate should be about 2–3 inches (5–7.5 cm).

Special soil can be used to grow plants (see the chapter on plants, pages 262).

Left:
Rinse the substrate to eliminate the fine particles that can make the water cloudy.

Below:
Spread the substrate on the tank base. A slight backward slope increases the apparent depth of field.

The various types of freshwater substrates

Fine sand from a riverbed is well suited to fish that live at the bottom of the tank. Its round grains are not too compact and can be penetrated by water and plant roots. Quartz sand is very popular: it is a little coarser and its grains are more angular. Gravel in various grades is widely available and enables aquarists to reconstruct a riverbed. Fine, black pebbles are sometimes used to create a distinctive look and set off colorful fish, such as neon tetras.

Aquarium kits

1 Natural forms of wood can create a lifelike underwater environment. As with rocks, it is advisable to buy wood specifically intended for aquariums rather than collecting pieces from the wild.

2 The best way to start an aquarium hobby is to buy one of the fully equipped tanks on the market. Manufacturers now offer not only an integrated filter system but also all the equipment required to install and run an aquarium without any setbacks.

3 Fluorescent bulbs are ideal light sources to support the growth of plants in freshwater aquariums and corals in saltwater tanks. Manufacturers offer a huge array of bulbs, with spectrums that have been meticulously formulated to supply the right type of illumination for both fish and plants.

4 Rocks are indispensable for decorating an aquarium. Aquarium stores sell rocks that have no effect on the water's physical and chemical qualities. This is not necessarily the case with rocks that you may be tempted to pick up in the countryside, as these may contain limestone or metallic ores.

5 The choice of substrate is important. They are available in a range of materials, grades and colors to suit all types of aquarium displays. Substrates enriched with fertilizer will boost plant growth, but should be used with care to avoid algal blooms.

6 A thermometer placed in a corner of the aquarium makes it possible to see at a glance whether the water temperature lies within acceptable limits. A defective water heater can lead to a number of problems in an aquarium.

1

2

3

6

4

5

Heating for the freshwater aquarium

This is essential for tropical fish, to ensure that the temperature remains in the range of 72–79°F (22–26°C). Aquarium stores sell easy-to-use electric immersion heaters with a heating element and thermostat built into a waterproof casing.

Right:
A thermometer (placed a reasonable distance from the heating system) makes it possible to check that the water temperature remains stable. Slight variations between the daytime and nighttime temperatures will not harm fish.

Below:
The required water temperature can be selected by adjusting the setting control on the thermostat.

Combined heater-thermostats (or heaterstats) reliably and effectively heat the aquarium water. They are available in a range of sizes and heat outputs. Allow about 4 watts per gallon (1 watt per liter), i.e., you would install a 100-watt heater in a 25-gallon (100 L) aquarium.

Heaterstats are attached to the aquarium glass with suction cups. They should be placed close to a filter or water current so that the heat they give off is evenly distributed. Nothing should touch the heating element and the thermostat control must be easily accessible. Do not leave heaterstats plugged into the power supply when the tank is empty. To monitor the temperature, install a thermometer – either a glass one containing alcohol (inside the tank) or one with a flat liquid crystal display (outside the tank). Position the thermometer on the opposite side of the aquarium from the heating system, to avoid biased readings. The thermometer is often set in a rear corner for cosmetic reasons, but make sure that it is easily visible so that you can check the water temperature every day.

Right:
Place the suction cups on the heaterstat.

Installing the heaterstat

The heaterstat is often placed in the corner of a tank to make it easier to hide.

Attach the heaterstat at an angle, with the thermostat on top and the heating element just above the substrate.

The circulation of water generated by the filter distributes the heat and establishes an even temperature throughout the tank.

Filtration for the freshwater aquarium

There are several types of filters: internal, external or undergravel. In most cases, they perform a double role: mechanical filtration to strain out solid particles, and biological filtration to deal with toxic organic wastes.

Internal filters

There are several types of internal filters but the principle is always the same. An electric water pump draws the water through a filter medium, generally a piece of foam. This performs mechanical filtration and also allows growth of the bacteria required for biological filtration. These filters are completely submersible, but always make sure that they are easily accessible so that the filter medium can be cleaned and changed as required.

Internal filters are ideal for small aquariums. The hourly flow rate must be equal to the amount of water that has to be filtered (i.e., a rate of 25 gallons (95 L) per hour for a 25-gallon (95 L) tank). They are generally installed with the help of suction cups in the corner of a tank; the water outlet must break the surface and lie diagonally across the tank. These filters should not be switched on until they are completely underwater.

In some models of internal filters, an air tube is connected to the water outlet. Air is drawn in from above the surface (by the Venturi effect), creating a stream of bubbles that help boost oxygen levels.

Internal filters

The watertight electric pump draws water through the filter medium.

In a filter of this size, the filter medium is usually a single block of foam.

The transparent plastic casing makes it easy to monitor the state of the filter medium.

Air is drawn in by the Venturi effect.

The foam can easily be removed and rinsed in a bowl of tank water when it becomes clogged.

This part houses a submersible pump and is attached to the glass by suction cups.

Water enters through a grill and exits through the top of the filter.

Installing an internal filter

Check with the instructions and position the filter head at or just below the water surface. If the outlet breaks the surface it will stir up the water and enhance oxygenation. Direct the outflow diagonally across the aquarium.

This internal filter fits into a cradle designed for a corner and is secured by suction cups.

Leave a space between the base of the filter and the substrate to prevent any debris interfering with water circulation.

Anatomy of an external filter

Filter wool traps fine particles.

Activated carbon removes toxins.

Filter wool separator.

Pellet biomedia supports bacteria.

Large-grade foam traps large pieces of debris.

Electric water pump with inlet and outlet for the water flow.

Internal compartment with filter materials.

Outer casing connected to the pump unit by clips.

External filters

External filters are ideal for large tanks. Providing both mechanical and biological filtration, they have two main advantages: they do not take up any space in the aquarium and they are easy to maintain – although they are more expensive than other filtration systems.

External filters come in many shapes and sizes. The positioning of the water inlet and outlet varies according to the model. The flow should be sufficient to filter the water twice in one hour, so a 50-gallon (190 L) tank requires a filter capable of treating 100 gallons (380 L) per hour. In practice, however, the flow will not correspond to the filter's capacity, especially if the filter media become clogged.

There are also different types of external filter media. They take up more space than those of internal filters, and they also enhance the growth of bacteria. The water first passes through a medium designed to trap large particles, such as large-grade foam. Next comes a medium designed to filter smaller particles and encourage the growth of beneficial bacteria – often sintered glass cylinders or similar "biomedium" with a large surface area. The final medium is even denser and retrieves even the smallest particles. The latter two media can be separated by a layer of activated carbon, which removes specific toxic substances.

The aquarium water is drawn through a strainer just above the substrate that prevents very large pieces of debris from being sucked up. The water runs through a wide, flexible tube into the filter canister. After passing through the layers of filter media, the filtered water passes along the return tube and back into the tank on the opposite side to that of the intake, if possible close to the surface for better oxygenation. Taps can be used to cut off the water supply during maintenance.

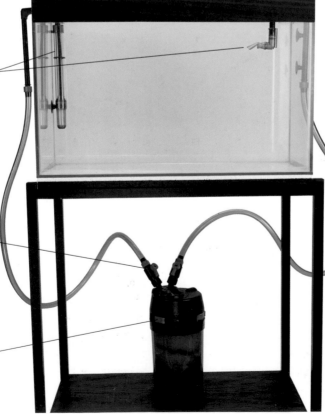

The intake (left) and return pipes (right) are installed on opposite sides of the tank.

Taps serve to switch off the water supply (after first switching off the pump) so that the filter media can be changed or rinsed.

If the filter is enclosed in a cabinet, make sure that it is well ventilated to avoid over-heating the water pump.

Undergravel filters

These consist of a perforated plastic plate or several small sections that clip together to fit tanks of different shapes and sizes. One or possibly two uplift tubes are set in the corners of the plate. The undergravel filter

Advantages and disadvantages of an undergravel filter

The main advantage is its low cost, even though it requires the acquisition of an air pump, or power-head. It is also a system that clarifies the water very quickly. The disadvantages are more numerous. First, it cannot be installed once the aquarium is in operation. Also, the substrate gradually accumulates debris before finally clogging up the system. This normally occurs after a few months, depending on the size of the substrate particles and the quantity of suspended matter in the water. If the water stops flowing even momentarily, the oxygen levels will plummet and the beneficial bacteria will die. There is a danger of the substrate fermenting.

Finally, drawing water down through the substrate is not beneficial to rooted aquatic plants. Nevertheless, an undergravel filter system can be useful for small aquariums or a breeding tank.

Fit the undergravel filter on the base of the aquarium Buy one that fits the tank exactly or use smaller modules.

Fit an uplift tube in one or both rear corners of the plate. Introduce an airline into each tube or fit a powerhead pump on top to set up the water flow.

Cover the filter plate with a layer of thoroughly washed substrate, about an inch thick.

plate is placed on the bottom of the aquarium, but it is raised slightly to allow water to pass underneath. The filter plate is covered with a suitable substrate that should not be too fine, in order to allow the water to circulate efficiently. In an air-operated system, an air line fitted with an airstone is fed down each uplift tube, or else the air line is connected directly to the base of the tube. Once the air pump is running, the rising bubbles draw water up the tube and create a flow of water down through the substrate to take its place. In a powered undergravel filter (UG filter) system, a small water pump (powerhead) is installed on top of each uplift tube, and this actively draws up water to set up the same pattern of circulation. The working principle of a UG filter system is that the substrate performs both mechanical and biological filtration.

Air pumps

There are several models on the market, in various forms and different airflows. Even though air pumps consume little electricity, they are powerful enough to operate several undergravel filters or airstones (for aerating the water), or a combination of the two. The most popular models work on a simple principle: a vibrator activates a diaphragm to create the airflow. Flexible, narrow plastic air line carries the pumped air to the aquarium.

Wooden decor for the freshwater aquarium

Wood is becoming increasingly popular in freshwater aquariums and is readily available in pet and fish supply stores. It is not recommended to use pieces gathered in the wild, as you could introduce various undesirable organisms (fungi, moss, lichens, insects, etc.) or drastically change the quality of the water (with resinous wood, for example).

Above:
A small, stiff brush should be used to carefully remove any foreign objects from pieces of wood and roots. Brush them first when they are dry, and then under running water.

Wood must be washed and brushed before use. Roots should be soaked in water (and may need to be held down to prevent them from floating) in order to eliminate their tannins, which are acidic. This soaking water must be changed every day, and the roots can only be put into the aquarium once this process is complete and the water is clear.

The wood and roots can then be pushed into the substrate and, if they are likely to float upward, wedged in place. While the tank is still empty, you can try out different positions for these decorative elements: vertical, horizontal or diagonal.

Placing wood in the aquarium

Try out several arrangements to find the most attractive configuration of the various pieces of wood.

Avoid blocking the filter's intake and outflow, and make sure that the wood does not touch the heating system.

Types of wood

Aquarists tend to prefer roots, branches and small pieces of wood collected from peat bogs, where they have often lain for several centuries. When submerged, this wood gives off a wealth of acidic substances that stain the water (suitable for some fish but not for most aquariums). Mopani roots are knotty, heavy and pale-colored. Mangrove roots display an array of gnarled forms that are guaranteed to create a visual impact. Cork can be used to simulate bark, but it must first be thoroughly soaked and washed. It can then be adhered to a rock with silicone sealant. Once the rock is lodged in the substrate, the cork is weighted down and cannot float. Cork can also serve to camouflage equipment or decorate the side and rear panes of the tank.

Rocky decor for the freshwater aquarium

Do not imagine that you can pick up any interesting-looking rock you may stumble across and put it in your aquarium. You must first check that it does not contain any substances liable to alter the quality of the water.

It is not advisable to mix different colors of rocks, especially if you are using wood as well; one single kind is sufficient to create an attractive decor. Rocks must be carefully brushed under running water before use. Fragments of different shapes and sizes can be assembled to recreate a natural setting, such as a grotto or mound. It is then essential to adhere the various elements to each other with silicone sealant. This will prevent any collapse caused by maintenance operations or even by the fish themselves, as some fish are more powerful than you think. All rocks, whether isolated or stuck together, must be placed on the substrate and then gently embedded in it until they touch the

Some suitable rocks

Metamorphic rocks such as schist are often brown, verging on red, which creates a strong visual effect. Slate is a popular choice, as its dark gray or black surface contrasts sharply with the rest of the decor, plants and fish. It can be found in thin sheets that are useful for hiding equipment without taking up too much room in the tank. Of the volcanic rocks, granite is the best known, but it is heavy  and difficult to break up. Lava and basalt (which are reddish or black) are more manageable. Aquarium stores also sell siliceous (petrified) wood. This is not really a rock but wood that has gradually been replaced with silica over time. Its rather pale color adds a striking, decorative touch.

glass that forms the bottom of the tank. It is best to set bigger rocks toward the back of the aquarium, medium-sized ones on the sides and the smallest ones in the front. Avoid knocking or scratching the glass panes when you put rocks in the tank.

Rocks to avoid

Rocks containing limestone are not suitable, except in some specific circumstances, as when you may need to increase the hardness and pH of the water. To find out whether a rock contains limestone, just splash a few drops of vinegar onto it: the chemical reaction will produce a few lively bubbles.

You must also avoid rocks with metallic ores. These can be identified by their metallic gleam when moved in the light.

Placing rocks in the aquarium

Rocks must be embedded in the substrate to ensure that they remain stable. They can also be stuck together.

Getting the freshwater system up and running

At this point, a simple checkup is called for. Is the equipment firmly in place? Is the decor stable and in the right position? Now is the time to make any adjustments, before the tank is filled – it will be much more difficult later on.

Filling the tank with water

Once the equipment has been set up, you can fill the aquarium. The simplest way of doing this is to use a large jug. Before adding any tap water, be sure to treat it with a dechlorinator or, at the very least, let it stand for 24 hours with an airstone running to dissipate the chlorine. Always pour the water gently onto the rocks to avoid dislodging the decor. If the aquarium has no rocks, place a small bowl on the substrate and allow the water to run into it. Stop filling the tank when the water reaches 2 inches (5 cm) below the desired level. It is perfectly normal for the water to be slightly cloudy at this stage.

Filling the tank
The receptacle used to fill the aquarium must be spotless and free of any traces of detergent.

Pour the water in gently to avoid altering the contours of the substrate.

Even if the substrate has been washed thoroughly, it is normal for the water to be slightly clouded by tiny particles in suspension.

Switching on the system
Stop filling the water a couple of inches below the normal water level.

Make sure that the heaterstat is fully submerged before switching it on.

The slight cloudiness of the water will disappear as the filtration starts to take effect.

Turn on the filter and adjust the angle of its outflow jet.

Switching on the power

It is now time to turn on the electrical equipment. Set the heater's thermostat at the desired temperature, usually about 77°F (25°C) for tropical fish, and switch it on. Monitor the internal or external filters, checking that the water is well aerated and, if necessary, adjusting the direction of the outflow jet.

The following day, make sure that the temperature has reached the desired level and that the filter is working properly. You can now plant the aquarium, following the advice in the chapter on plants (see pages 256–81), and then proceed to the lighting. Finally, you can then fill the tank right up to its final level.

Lighting

The lighting, which may be integrated into the hood or suspended above the tank, can now be switched on. You can use a small electric timer to schedule the light levels.

Do not forget to place a sheet of glass or highly transparent plastic between the water and the hood, to prevent any condensation reaching the fluorescent bulbs (even if these satisfy safety regulations and their ends are sealed).

You must now wait patiently before putting the majority of the fish, particularly any large ones, into the aquarium: for as long as the filter system needs to reach maturity – up to three weeks.

Planting

After planting, top up with water to the required level: this will be hidden behind the decorative black strip.

This is the time to plant the aquarium – before the tank is full.

A plastic or glass sheet prevents any evaporation.

Lighting

The hood, with the lighting system inside, is then placed on top of the aquarium. All the electrical fixtures are waterproof.

Once the lighting is switched on, you can finally appreciate the decor in all its glory.

Introducing fish into a freshwater aquarium

Once an aquarium has been planted and all the equipment is working smoothly, you must wait before putting the majority of the fish into the tank.

All varieties of goldfish, red butterfly moors are pictures here, require a fairly spacious aquarium.

Why wait after planting?

First of all, the aquarium needs to become balanced. The nitrogen cycle must establish itself, and this will require at least three weeks. However, there must be at least a couple of fish in the tank producing waste to get the system going. A hardy, inexpensive algae-eater, such as *otocinclus affinis*, is a good first occupant for a freshwater tank. During this period, the beneficial bacteria required to "power" this cycle colonize the filter media.

At the same time, the plants take hold, avoiding any possible accidental uprooting by fish traumatized by their introduction into the tank. Some plants may even start to grow.

This waiting period will allow you to reflect on your choice of fish. Before making your selection be sure to look in books such as this one for helpful suggestions.

Timetable for the introduction of fish

It is always preferable to put the fish into the tank in several batches, rather than all at the same time.

Some species can be introduced into their new home about three weeks after the planting. Start off with the smallest and shyest specimens, so that they will have time to grow accustomed to their new environment before the arrival of larger or friskier fish, about two weeks later. Also, larger fish put more strain on the system so should be introduced last. Any remaining fish can be added a week afterward.

The bloodfin tetra (*Aphyocharax anisitsi*) thrives as shoals in water at about 72°F (22°C).

Left:
The red-tailed black shark *(Epalzeorhynchos bicolor)* is fairly sociable and accepts the presence of other species, but will be aggressive with its own kind – one only per tank and do not mix it with similarly colored or shaped tankmates.

Below:
The lionhead cichlid *(Steatocranus casuarius)* can be aggressive if provoked to defend its territory.

How many fish in an aquarium?

The maximum number of fish that can be put in a tank depends on the amount of space available.

Many novices tend to overcrowd their aquarium, but this can upset the balance. There are a number of rules of thumb. The following is based on the calculation of the aquarium's surface area: allow 12 square inches (75 sq. cm) per 1 inch (2.5 cm) of fish length. Remember to consider the maximum size that a fish can reach in captivity. The caudal fin is generally excluded from the equation, particularly in fish where it is exceptionally long (either naturally or through special breeding techniques).

How to choose fish

It is always advisable to start with robust, well-known species. There is a huge range available, and the final decision depends on both personal taste and budget.

It is worth choosing not only open-water fish but also bottom-dwellers and surface-swimmers in order to have all the zones in the aquarium occupied. Of course, these species must be compatible.

A common mistake is to introduce many species, each represented by one or two fish. Although some fish can thrive on their own, others prefer to live in pairs or in shoals, and their way of life must be respected. In a 25-gallon (95 L) tank, a maximum of four or five species is reasonable.

MAXIMUM SIZE OF FISH

Large fish will not be at ease in a small aquarium, especially if they are strong swimmers. There are therefore limits that should not be exceeded. Here are a few examples:

Dimensions of tank (length x width x depth)	Tank volume	Maximum adult length	Maximum number
24 x 12 x 12 in. (60 x 30 x 30 cm)	14 gal. (54 L)	2–3 in. (5–8 cm)	10
32 x 12–14 x 16 in. (80 x 30–35 x 40 cm)	25–30 gal. (95–110 L)	4 in. (10 cm)	8–9
40 x 16 x 16 in. (100 x 40 x 40 cm)	45 gal. (170 L)	6 in. (15 cm)	8–9

MAXIMUM NUMBER OF FISH PER TANK

Tank dimensions (length x width)	Tank surface area	Max. total length for tank size	Max. no. of adult fish 2 in. (5 cm) long
24 x 12 in. (60 x 30 cm)	288 in.² (1,800 cm²)	24 in. (60 cm)	12
32 x 12–14 in. (80 x 30–35 cm)	384–435 in.² (2,400–2,800 cm²)	32–36 in. (80–90 cm)	16–18
40 x 16 in. (100 x 40 cm)	640 in.² (4,000 cm²)	52 in. (130 cm)	26
48 x 16–18 in. (120 x 40–45 cm)	768–864 in.² (4,800–5,400 cm²)	64–72 in. (160–180 cm)	32–36
60 x 20 in. (150 x 50 cm)	1,200 in.² (7,500 cm²)	100 in. (250 cm)	50

Adding the fish

1 Fish can be introduced into the aquarium once it has become balanced, about three weeks after the addition of the water. The time has come for the eagerly awaited trip to the pet store, to choose fish and plunge into the adventure of looking after an aquarium.

2 Place the bag containing the fish in the aquarium. Allow the bag to float for about 30 minutes for best results. This serves to balance the temperature of the water in the bag with that of the aquarium.

3 It is inadvisable to try observing whether everything is going well during this period. Handling the bag will be a source of stress for the fish.

4 Open the bag very carefully by cutting the plastic just under the knot with a pair of scissors. Put a little of the aquarium water into the bag, using a food-safe container. It is worth repeating this operation about 10 times over the course of about 10 minutes. To achieve a good acclimatization, the volume of water added should be twice the initial volume in the bag.

5 Lift the fish gently from the bag with a net. This avoids introducing water from the dealer's aquarium into the aquarium, as it is a possible source of contamination.

6 Before releasing the fish into the aquarium, take a close look at them to ensure there are no signs of a problem. If in doubt, return any questionable fish to the bag and take them straight back to the store.

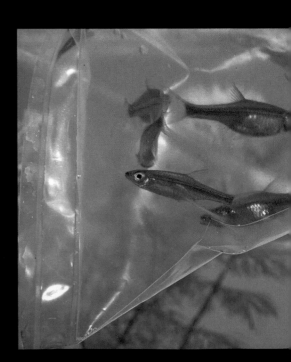

Where to obtain fish

Pet stores and aquarium dealers are the traditional options. There is a wide range of species, so it is a good idea to visit several stores to see what fish are on the market. Aquarium dealers can usually be relied upon for useful advice.

You can also obtain fish through friends or an aquarium club (it is strongly recommend that you join one, in any case). Old hands are only too happy to share their experiences and may even offer some of the hobby fish they have bred themselves.

long, it is wise to provide insulation, in the form of a layer of newspaper or, preferably, a Styrofoam box.

Transferring fish to the aquarium

If the trip home has not exceeded 20 to 30 minutes, float the bag containing the fish on the surface of the tank (watch out for spills). This allows the temperature inside the bag to adjust to that of the aquarium, thereby preventing a vast change in temperature. Let the bag float for about half an hour.

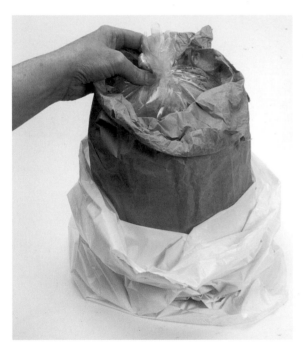

Above right:
The bag must float for 30 minutes to reach the same temperature as that of the tank.

Above:
A transportation bag for aquarium fish must contain two parts air and one part water. It can be wrapped in newspaper to prevent heat loss. A brown paper bag will keep the fish calm in subdued light.

Transportation

Aquarium retailers put fish in plastic bags, although some hobbyists prefer plastic bottles with a wide opening. What is essential, however, is that the fish have sufficient oxygen during transportation. The water level must not exceed, under any circumstances, one-third of the volume of the bag, with the other two-thirds reserved for air. If you are transporting more than a few fish, use several bags. These must be slanted during transportation, as this enhances the oxygenation. If the journey home is short, heat loss is minimal and a thermally insulated bag will not be required. On trips more than an hour

If the journey home has taken longer than 30 minutes, open the bag to reoxygenate the water. Then float it in the

tank, securely anchoring it to the side to stop it from tipping over, for the same period of time.

Next, gently remove the fish from the bag with a net and put them into the aquarium. In order to avoid increasing the stress that the fish have already experienced during their journey and change of water, some aquarists keep the tank lights turned off during this process.

The most courageous fish may explore their new territory right away. Expect to wait at least 24 hours, however, before you catch sight of them (it is not worth feeding them during this time). They will gradually start to reappear, acclimatize and get to know each other.

An aquarium in full operation

When fish are first put in an aquarium, they often hide in the plants or behind the rocks.

Check that the hood is tightly shut to prevent traumatized fish from jumping out of the tank.

The most courageous and least stressed fish will not wait very long to explore their new environment.

From this moment on, you must regularly monitor the temperature, bearing in mind that slight variations are normal.

Maintaining a freshwater aquarium

In a closed setting like an aquarium, regular maintenance is vital to sustain a healthy environment for your fish. This is not taxing – it takes an average of one hour per week. Some operations must be carried out every week, every two weeks or every month, and others less regularly. It is helpful to use a small notebook to schedule the maintenance program of your aquarium, and also to jot down all your observations of the plants and animals.

Temperature and quality of the water

Check the temperature every day, at the same time as you feed the fish. When you are familiar with your setup, just brushing the glass with your hand will tell you if the tank feels "right." It is advisable to check the levels of nitrites and ammonia, as well as the pH, about twice a week. The sooner you find any abnormal values the greater opportunity you have to remedy the problem before your livestock are threatened.

Changing the water

This is absolutely essential, even if the water is well filtered. It is normal for certain substances to accumulate over time, particularly nitrates. This often leads to a coloring of the water, which is clearly visible if a white object is put into the tank (it takes on a yellowish appearance). Generally speaking, it is sufficient to change 10 percent of the water volume in a balanced aquarium once a week. The replacement water must have approximately the same characteristics as the water that has been siphoned off (temperature, pH and hardness).

It is a good idea to carry out a partial water change after vacuuming the substrate and cleaning the inner glass surfaces. This will help to remove floating debris.

Some parameters of the water need to be monitored regularly: pH, ammonia, nitrites and nitrates. The tests on the market make this easy to do.

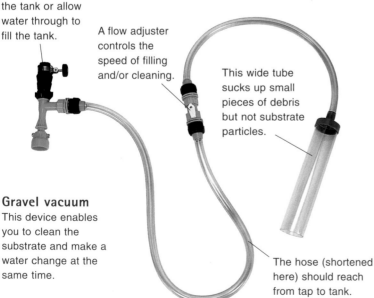

The tap pump can either create suction to clean the tank or allow water through to fill the tank.

A flow adjuster controls the speed of filling and/or cleaning.

This wide tube sucks up small pieces of debris but not substrate particles.

Gravel vacuum
This device enables you to clean the substrate and make a water change at the same time.

The hose (shortened here) should reach from tap to tank.

What to do in the event of a power failure

There is no problem if it is short. The temperature of the aquarium will go down slowly for the first hour (particularly if it is a large aquarium), without any serious effects on the fish and plants. If it looks as if the power failure will last longer, you can wrap the tank in a blanket to slow down any heat loss. It is also possible to heat up the aquarium by floating bottles of hot water on the surface.

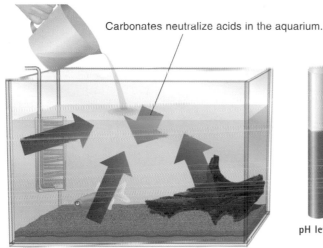

Carbonates neutralize acids in the aquarium.

pH level

Regular water changes help to replenish carbonates, which prevent any extreme variations in the pH (buffering effect).

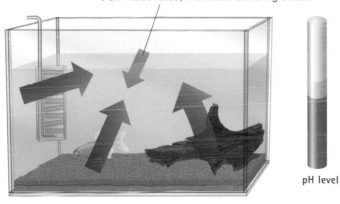

Few carbonates, moderate buffering effect.

pH level

If the water is not changed regularly, or if it is not very hard, some of the acids will not be neutralized and the pH will drop.

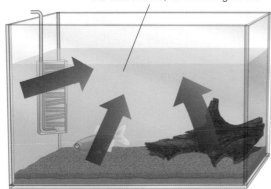

No carbonates, no buffering effect.

pH level

The acids normally produced by chemical and biological processes are not entirely neutralized and the pH plunges.

What to do if the water is too hot

This can occur in summer, during a heat-wave. The thermostat will obviously turn off the heating, but the increase in room temperature can have a drastic effect on an aquarium. There is no danger up to about 80°F (27°C), but after that the water must be cooled down. To do this, use ice cubes in a plastic bag or a plastic bottle half-filled with melted ice from the freezer and topped up with tap water; allow this to float in the tank until the temperature drops. You can buy coolers of various designs that will chill the aquarium water. These work well but are relatively expensive to buy.

The substrate

Despite the presence of a filter, some detritus will still accumulate on the substrate: fish excrement, plant debris and uneaten food. You can siphon these off with a hose leading to a sink, toilet or garden, but it is more practical to use a siphon-action gravel vacuum. Simply follow the instructions to start up a siphon flow and use the wide plastic tube to disturb the substrate so that light debris is whisked away in the water flow.

Other devices are available for cleaning the substrate. This one is easy to use, runs on batteries and does not need to be connected to a tap. You can also buy a miniature air-powered vacuum cleaner that collects debris in a mesh bag.

Maintaining the filter

If the filter is not maintained, less water will flow through it. The filter media can get clogged with solid particles filtered from the water, as well as various organic debris produced by the livestock. These toxic substances sometimes give rise to a noticeable putrid smell.

Removing the filter from the aquarium
Switch it off and detach it from its support or the side of the aquarium. This photo shows a filter that is very dirty, especially around the water intake grill.

Therefore, it is vital to clean the filter media, even though this will also remove the beneficial bacteria involved in the nitrogen cycle and, therefore, the aquarium's biological filtration process. So, only half the filter media should be cleaned at any one time; the cleaned media will quickly be recolonized by the bacteria from the unwashed half, which you can clean the following week, and so on.

In order to clean a piece of foam, squeeze it several times, as if it were a sponge, under warm, running water. An aquarist with several tanks can use a washing machine to clean several pieces of foam at the same time, but only with "rinse" and "spin" settings. No type of detergent should be used, under any circumstances.

If you are cleaning the whole filter medium, it is a good idea to rinse it in a bowl of tank water so the bacteria are not affected. Finally, thoroughly brush the intake grill of internal and external filters.

The pump housing

This must be thoroughly washed and rinsed, and the water intake grills and any prefiltration equipment cleaned of all debris.

Dismantling
Separate the pump from the filtration compartment over a large bowl filled with tank water, and insert the compartment along with the filter equipment.

Air pumps

Air pumps are sometimes fitted with a small filter near the air intake. Be sure to change this regularly, along with the diaphragm, following the manufacturer's instructions.

Cleaning the aquarium glass

The aquarium glass is often covered with microscopic algae. This is not necessarily a sign of poor water quality – in fact, it can indicate that the aquarium is well balanced. Nevertheless, it is unattractive and the glass must be cleaned. There are several methods for this. Aquarium stores sell scrapers of various designs that are very effective. In two-part magnetic scrapers, one part has an abrasive inner surface that rubs the inside surface of the glass, and is moved by the other part magnetically "attached" to it on the outside. Glass in tanks with a relatively low water level can be wiped with a slightly rough shower glove, kept exclusively for this purpose. When cleaning the glass, it is usually impossible to avoid leaving particles suspended in the water or on the substrate. This is why it is always preferable to clean the glass just before vacuuming the substrate.

There are various types of tools for removing algae from the aquarium glass. *Above:* This algae scraper with a handle is easy to use from above the aquarium. *Left:* An algae magnet allows you to clean the glass without getting your hands wet.

Cleaning the foam

Squeeze the foam gently in a bowl of tank water to eliminate any small pieces of debris. If possible, clean only half of the filter media at a time to preserve some of the beneficial bacteria.

Reassembling the filter

Once the cleaning is finished, replace the foam, put the filter back together again and reinsert it in the aquarium. Make sure that it works properly once it is switched on.

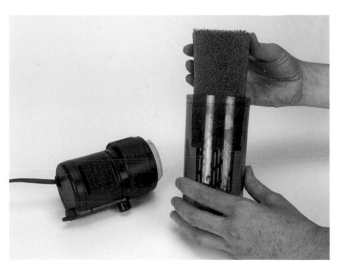

Right:
Undesirable algae sometimes grow so fast that they threaten to overrun the aquarium. You must try to eliminate them before they reach that stage, but it is often difficult to be totally free of them.

Above:
If algae start to grow on the leaves of plants, they must be cut off.

Far right:
To remove algae from rocks or pieces of wood, take these items out of the tank and brush them under running water.

Lighting

Fluorescent bulbs must be changed at least once a year, but some aquarists prefer to do it every six months. The difference in the intensity of old and new bulbs is surprising, and illustrates why they need to be changed so often. If several bulbs are used, do not change them all at once, to avoid any radical difference in the light falling on the plants.

The fight against algae

Algae can grow in both a recently installed aquarium and a tank that has been in operation for several months. They are desirable in small quantities (algae can even be considered beneficial and an indicator of good balance in an aquarium), but if their growth is left unchecked they can spring a few surprises. As well, the visual effect is unsightly.

Restricting the growth of algae

There are some preventive measures that are easy to put into practice. First, you can introduce one or two herbivorous fish into the aquarium, such as the Siamese algae-eater (*Crossocheilus siamensis*), shown below right, or a catfish from the *Ancistrus* or *Otocinclus* genus. Next, you should control the quality of the water by partially changing it on a regular basis. The lighting must be neither too strong nor too weak, and any plant fertilizers (if used) must be dosed correctly, in accordance with the manufacturer's instructions. Do not forget that algae have less chance of growing in a profusely planted aquarium.

Green water is caused by microscopic algae (the vegetable equivalent of plankton). They grow as a result of overly strong lighting, so light levels should be reduced; at the same time, make partial water changes.

Brown algae can appear on the glass or the decor, often due to excessively weak light, particularly in freshwater. In this situation, intensify the lighting and use a slightly softened water, after first cleaning the glass.

Filamentous algae are a greater nuisance. They grow quickly and are very difficult to eliminate totally. They grow on decor, but also on other plants, which end up being stifled. An overproliferation of filamentous

algae is generally caused by lighting that is too strong and water that is rich in organic matter and nitrates. Most of these algae will have to be removed by hand – for example, by winding the filaments around a stick. You can also try taking out some pieces of the decor and brushing them thoroughly under water. It is sometimes even necessary to cut off the leaves of the plants that are most affected. Finally, you can use algicide products available in aquarium stores.

In all cases of algae growth, it is vital to siphon off any that remains on the substrate after your attempts to eliminate it.

MAINTENANCE SCHEDULE FOR A FRESHWATER AQUARIUM

	Fish	Plants	Water	Substrate	Equipment
Daily	Feed them; observe their behavior and check for possible signs of illness. Remove any dead fish.		Monitor the temperature.		Make sure that the filter (and/or pump) is working properly.
Twice a week			Monitor the pH and levels of ammonia and nitrites.		
Once a week		Remove dead or fallen leaves.	Change 10 percent of the water volume.	Siphon off waste with the water change in the first month after the aquarium has been set up.	Clean half the filter media, then the other half the following week. Clean the outside glass.
Every 2 weeks		If necessary, prune fast-growing plants.		Siphon off the waste at the same time as the water change.	Clean the inside front glass and the condensation tray or cover glass.
Every 3 to 6 months		If necessary, propagate slow-growing plants			Change any activated carbon in the filter. Service the air pump and filter motor, following the manufacturer's instructions.
Once a year					Change the fluorescent bulbs.
When necessary		Remove undesirable algae.	Readjust the water level.		Replace the filter media.

The ground rules of maintenance

Change the fluorescent bulbs once a year.

Clean the condensation tray or cover glass every two weeks.

Carry out partial water changes on a regular basis.

Check every day that the electrical equipment is working properly.

Clean the front glass to remove algae.

Look after the vegetation and remove dead leaves.

Inspect the fish every day.

Clean the filter media regularly.

Service the filter motor according to the maker's instructions.

Check the temperature every day.

Remove uneaten food every day.

Siphon off waste from the substrate.

Equipment for a saltwater aquarium

Although some elements are the same in both freshwater and saltwater aquariums (the heating, for example), specific equipment is required in a saltwater tank.

The tank

The water in a saltwater aquarium must always be impeccable, and a large tank is more capable of providing stable conditions over a period of several months. Also, fish are accustomed to far more space in a natural saltwater setting than they are in many freshwater habitats.

The bigger the tank, therefore, the greater the chances of success, as the fish will be able to live in more appropriate conditions. It is pointless to attempt keeping fish in good conditions if the volume of the tank is less than 40 gallons (150 L), which corresponds to a tank length of about 39 inches (1 m).

Obviously, small species can be kept in a tank of such modest dimensions, but big fish need an aquarium at least twice this size, especially if it is also to contain invertebrates.

The background

As in a freshwater tank, you must remember to add a background while you still have access to the rear of the aquarium. The most popular backdrops used in saltwater tanks are uniform blue or black.

The substrate

The substrate plays a very different role as that in a freshwater aquarium. There are no rooted plants in the sea (except seaweed), so the substrate does not need to be deep or rich in mineral elements.

A depth of ¹/₂–1 inch (1.25–2.5 cm) is sufficient

Rinse the substrate in several changes of clean water in a light-colored bucket.

(apart from rare exceptions). A deeper bed increases the a risk of deoxygenation, which triggers the production of toxic gases.

Make sure that the substrate is washed until the rinsing water is clear before putting it on the base of the aquarium.

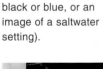

As in a freshwater tank, a background may be attached to the rear glass (plain black or blue, or an image of a saltwater setting).

Materials suitable for a saltwater substrate

Coral skeletons, roughly ground into coral sand, is often found in saltwater aquariums.

Another option is marl, which is found on some seabeds; it is made up of calcareous algae and seashells and has a coarser texture.

Both these materials contain calcium, so they prevent any significant variations in the pH (which must remain at about 8.3).

Left:
Some fish, such as this yellow-headed jawfish *(Opistognathus aurifrons)*, need a substrate of some depth so that they can hollow out a burrow.

Below:
For the vast majority of species, including this lionfish *(Pterois volitans)*, a depth of $1/2$–1 in. (1.25–2.5 cm) is sufficient.

Live sand

Live sand is a substrate that contains bacteria (and other organisms) that participate in the nitrogen cycle. It can be placed in an aquarium during the initial installation or once the tank is already in operation.

Live sand is available from aquatic retailers, but it can also be taken from a tank that has been balanced for a few months (as from a friend or aquarium club). In this case, it should make up 10 percent of the total volume of the substrate in the recipient aquarium.

Left:
Once it has been carefully washed, spread the substrate evenly over the base of the aquarium. The most commonly used substrate is coral sand, readily available at your aquarium dealer.

65

Heating

The principle here is the same as for a freshwater tank. Watertight submersible electric heaters are used, and the power required is calculated based on 4 watts per gallon (1 watt per liter) of water. In order to distribute heat evenly in a large tank, use two heating units with a combined power equal to the required total. This means that two heaters in a 75-gallon (or 300 L) tank should each be rated at 150 watts. They should be placed at opposite ends of the aquarium (in the rear corners, for example). A thermometer is essential for monitoring the temperature.

Power required for pumps and filters

Considerable power is needed for pumps and filters. Half the water flow goes through the filter, the other half through a submerged pump with no filter to provide strong water currents. In a fish-only aquarium, the total flow of water must be the equivalent of three to five times the volume of the tank per hour. A 75-gallon (285 L) tank, therefore, requires a pump capable of moving between 225 and 375 gallons (850 and 1,425 L) per hour; the proportion going through the filter would be 115–118 gallons (435–447 L) per hour.

In reef tanks with invertebrates, the flow must be even stronger. All the water must be circulated at least 10 times (or more) every hour. In a 135-gallon (510 L) tank, therefore, the pump should circulate 1,350 gallons (5,100 L) per hour, with a filtration system of the same power.

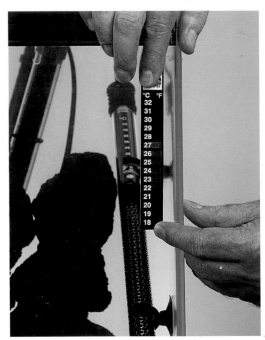

Left:
The heating units and thermometers used in saltwater tanks are identical to those found in freshwater aquariums. They are often placed in the corners of the aquarium.

Left:
Fit a guard (center left) to protect nervous fish from injuring themselves if they shelter too near a heater.

Right:
Two heaters can be used in a large salt-water aquarium, as long as their combined power equals the required total. Placing them at opposite ends of the tank ensures that the heat is evenly spread.

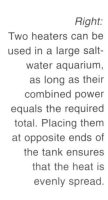

An external filter for a saltwater aquarium

These plastic tubes carry water to and from the aquarium.

Filter wool to trap fine particles.

Activated carbon removes organic toxins.

Filter floss as a separation layer.

Pelleted biomedia supports beneficial bacteria.

Foam to trap large particles.

Shut-off valves allow the filter to be disconnected for cleaning.

The electric water pump is housed in the upper part of the filter.

Water must pass continuously through the filter media. If the flow stops, the media will become anaerobic (without oxygen) and the toxic substances that form may harm the fish when the filter starts up again.

Circulating and filtering the water

The water in a saltwater tank must be extremely well oxygenated, and therefore thoroughly agitated. The water outflow from an external filter is used for this purpose. If this is not sufficient, a submerged pump with no filtration compartment can be used to provide essential circulation. Filtration operates on the same principle as in a freshwater tank, but a large external filter is generally used, with small internal filters sometimes used to complement. The filter media are the same as in freshwater systems.

This protein skimmer is placed outside the tank.

The skimmer is supplied with water from the aquarium by a submersible electric water pump.

Fluidized bed filter

This is an excellent external biological filter that must, however, be used as a complement to other systems. A current of water keeps sand in suspension inside a column. The sand is fairly fine and is almost entirely made up of silica. The filter provides a very large surface area on which the bacteria needed for the nitrogen cycle can flourish. This type of filter is very efficient for medium-sized tanks.

The protein skimmer

This filtration unit is not used in freshwater tanks but is indispensable in saltwater aquariums. A protein skimmer removes dissolved organic substances, particularly proteins, which other systems are unable to filter out. Their accumulation would risk upsetting the biological balance of the aquarium. Protein skimmer designs vary, but they all mix air bubbles with tank water to create a foam that rises to the top of the unit. Protein scum is carried upward in the foam and settles out as a yellow liquid when the foam collapses (see sidebar, at right). In the most advanced models, a powerful water pump equipped with an impeller "chops up" the incoming water and mixes it with air bubbles created by a Venturi-effect intake.

Filtration in a reef aquarium

A reef aquarium brings together corals, anemones or related species, as well as other invertebrates, such as shrimp. It is the dream of all saltwater aquarium enthusiasts. However, filtering such a tank requires a high degree of specialized knowledge. The most successful technique seems to be the Berlin method (so-called because it was perfected in a marine aquarium club in Berlin). This approach relies heavily on the biological filtration activities of live rock (see page 71) in conjunction with a protein skimmer to remove organic waste.

In this type of filter, the sand is kept suspended by the circulation of the water. This moving biological bed serves as a good support for the development of beneficial bacteria that process nitrogenous waste.

FLUIDIZED BED FILTER

The water flow into the filter is regulated by rotating this control.

Water from the aquarium flows in through this tube.

Water returns to the aquarium out this tube.

This cartridge contains activated carbon, which removes impurities from the water.

The medium used in this filter is special silica sand that provides a large surface area for beneficial bacteria to colonize.

An optional cartridge can be attached that removes phosphates from the water.

The principle of a protein skimmer

Protein skimming relies on the natural tendency for organic molecules (i.e., those in protein waste) to "stick" to the surface film around bubbles. Protein skimmers generate masses of bubbles and maximize the exposure of incoming tank water to this effect.

How a protein skimmer works

Protein waste collects in the upper chamber in the form of a yellowish foam. This settles out to a liquid that can be discarded.

Cleaned water returns to the aquarium from the outer cylinder of the skimmer.

A Venturi air intake creates a mass of bubbles in the water flow.

The water enters from the aquarium through the lower part of the skimmer.

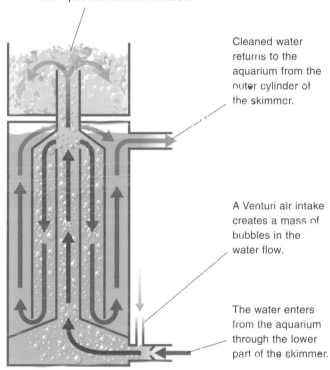

Connecting an external filter

The filtered water returns to the aquarium slightly below the surface via a rigid spraybar.

This is the intake pipe for water being drawn from the aquarium and passing to the external filter. The grill prevents large pieces of debris or fry from being sucked up into the filter.

Decor in a saltwater aquarium

Here, decor is indispensable. Fish need it to mark out their territory, and it also provides them with places to hide. Obviously, the various pieces making up the decor must not alter the characteristics of the water, so this rules out any wood and rocks containing metallic ores.

A number of other rocks can be used, particularly limestone, as it helps to maintain the pH. Rocks are available in aquarium stores, but it is also possible to collect them from seashores.

All solid materials must be carefully brushed and washed before being put into an aquarium. It is advisable to glue them together with aquarium silicone sealant to prevent the fish from causing landslides or

Limestone can be used in saltwater aquariums – in contrast with freshwater tanks – provided it does not contain any metallic ores. Animal shells, such as barnacle shells (below left), are also sometimes used.

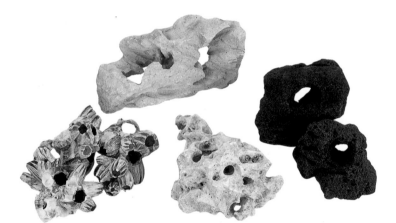

Coral as decor

The skeletons of hard corals, which are formed almost entirely of calcium, are sold by aquarium retailers. They create a striking visual impact, and come in such a diversity of forms that it is unusual to find any two alike. Their use in aquariums encounters stiff opposition, however, particularly from scientists eager to protect natural reef environments. Synthetic versions look surprisingly realistic.

rearranging their furniture! This approach will also enable you to build caves and rockscapes. Of course, you must ensure that the return pipes from the filter and pumps remain unobstructed.

Filling the aquarium and making the saltwater

The term "salt," used here, refers to the special aquarium salt mixes on the market. Fill the aquarium with water of very limited hardness (ideally, reverse-osmosis water). In order to avoid disturbing the decor or sending substrate into suspension, gently pour the water onto a rock.

As you will need to adjust the salt con-

Rocky decor must be totally stable. If you want to create a mound or provide caves and hideaways, adhere the different elements of decor together with aquarium silicone sealant.

centration and maybe adjust parts of the decor, stop filling the tank when the water is a couple of inches below what will be the final surface level.

Switch on the filter, pump and heating systems, then wait until the temperature reaches about 77°F (25°C).

Animal shells as decor

It is possible to use the shells of bivalve molluscs (which come in two parts) or of gastropods (a single shell, like that of a snail). The carapace of an acorn barnacle can also serve as natural decor. These shells must always be meticulously cleaned, especially if they were found at the seaside, as they may still contain fragments of flesh capable of polluting the water in an aquarium. Soak them for several days in several changes of clean water, then brush and rinse them thoroughly. Synthetic versions of animal shells – as well as a vast range of other seawater creatures and items – are available from aquarium suppliers.

Live rock

These pieces of coral rubble that form after storm damage in their native habitats are collected from tropical reefs and sold in aquarium stores. They contain live, often tiny organisms and bacteria that enhance the balance of an aquarium. Obviously, such rocks should not be scrubbed and cleaned like a normal rock; nevertheless it is preferable to keep live rocks in a quarantine tank before putting them into the display aquarium.

When filling the tank for the first time you can pour the salt in directly. It will be totally dissolved in less than 24 hours. Subsequently, never pour the salt into the tank, as you risk making disastrous errors. Instead, premix your saltwater in a large bucket. In order to estimate the amount of salt to add, calculate the volume of the aquarium and subtract 10 percent, which is the approximate volume taken up by the substrate and decor. The result will be the volume of water actually in the tank. The ratio of salt to water will be specified on the packaging.

Having calculated the amount of salt required, pour it directly into the tank (or into a bucket of RO water, after the aquarium is set up). Circulating the water through aeration and filtration speeds up the dissolving process, which will be complete in less than 24 hours.

71

Adjusting the density

A hydrometer sinks lower or floats higher to reflect specific gravity. In this case, it is extremely low.

Right:
If the specific gravity is too low, add salt, wait for an hour and then take another reading.

Below:
If the water is too salty, and the specific gravity therefore too high, add fresh RO water and taking a new reading.

Top up the tank with fresh reverse-osmosis water, wait for an hour until the salt is fully mixed into the water, then measure the specific gravity (SG), which is the ratio between the density of saltwater and pure water (which has a specific gravity of 1). The reading should be 1.023 at a temperature of 77°F (25°C).

If the specific gravity is too low, add more salt in small amounts until you achieve the desired value. If it is too high, siphon off a little salty water and replace it with fresh RO water to bring the SG down to 1.023.

Whether you add salt or water, it is important to wait for one hour before measuring the SG again.

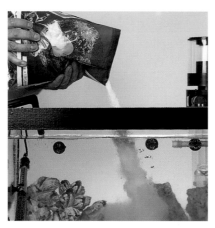

A colored strip on the hydrometer neck indicates the correct value. The thermometer also confirms the correct temperature for the reading.

Essential partners

Micro-algae known as zooxanthellae live in the tissues of some corals (as well as certain sponges and clams). Like other plant cells, they use light energy and carbon dioxide to produce simple sugars through photosynthesis. They also use nitrate and phosphate as fertilizers. Thus, in this partnership with their hosts (called endosymbiosis), they not only provide nutrients for the animal cells they live in, but also help to dispose of their waste products.

Lighting

In a modestly sized saltwater aquarium, with a depth of no more than 20 inches (50 cm) and stocked only with fish, fluorescent bulbs will be sufficient. Pink-colored bulbs, originally intended for plants, are not essential, unless seaweed is present. The most usual option is white "daylight" bulbs.

A saltwater aquarium may, however, also contain invertebrates, such as corals and anemones, and some of these live in symbiosis with microscopic algae (zooxanthellae) that derive their energy from light. For these, the lighting must be strong enough to stimulate their life processes.

In this case, fluorescent bulbs are not sufficiently powerful, so metal-halide lamps should be used. These lamps more closely simulate the intensity and "color" of natural light over a tropical reef (see sidebar, right). To complement white fluorescent bulbs and metal-halide lamps, it is common practice to add one or two blue actinic bulbs to the lighting above saltwater tanks. These bulbs not only provide the right spectrum of light for specific invertebrates, such as corals, but also, with suitable time switching, can produce a

realistic "dawn and dusk" effect at the start and finish of the illumination period. Whatever lighting is used, it will penetrate the tank more thoroughly if the water is very clear. This, of course, requires highly efficient filtration.

Metal-halide lamps

These are reserved for deep tanks, particularly those with coral.
They are more expensive but also more powerful than fluorescent bulbs. Ideally, allow 150–250 watts of metal-halide lighting for a water surface area of about 320 square inches (2,000 sq. cm). These lamps give off a great deal of heat and are suspended 12 inches (30 cm) above the aquarium in custom-made housings. In these circumstances, the tank no longer has a hood, but is covered with a sheet of glass to reduce evaporation. Metal-halide bulbs last for about a year.

A month of patience

Time is needed for an aquarium to find its balance and for bacteria to grow in the biological filter. The nitrogen cycle also takes time to establish itself – four weeks on average, but sometimes more. An organism to produce the nitrogen is needed to start the cycle, and the introduction of bacteria will help sow the filter. Live rock is a source of both benefical bacteria and nitrogen (see page 71). Monitor the levels of ammonia and nitrites until they reach zero. There will still be nitrates, however, but these can be reduced with reactors that encourage the growth of nitrate-reducing bacteria, or through regular, partial water changes.

Blue actinic bulbs complement white bulbs and metal-halide lamps.

You must wait a month or more before introducing any fish. This is the average time it takes the nitrogen cycle to establish itself.

Left:
White fluorescent and blue actinic bulbs are commonly used to illuminate saltwater aquariums.

Maintaining a saltwater aquarium

The guidelines already given for freshwater are equally applicable to saltwater, although some further checks and adjustments are vital.

Right:
To raise clownfish, such as this tomato clownfish *(Amphiprion frenatus)*, and their anemones, the water must be of impeccable quality.

Water quality
Monitor the levels of ammonia and nitrites, along with the pH, on a regular basis – around twice a week. As water evaporates from the aquarium surface, the specific gravity will increase slightly over time. To counteract this, top up the aquarium with freshwater; reverse-osmosis water is ideal.

Water changes
These must be undertaken with water of the same specific gravity as that in the aquarium.

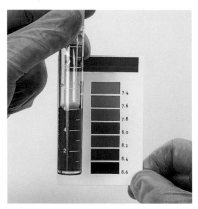

Above:
Regularly monitor the water in a saltwater aquarium regularly. This analysis of the pH shows that the level is correct at about 8.3.

Right:
Newly mixed saltwater added to the aquarium during water changes should be prepared in a large bucket, from freshwater of the highest quality – if possible, from a reverse-osmosis unit such as this.

You can prepare some in advance and store it in bottles or jars, if possible in a cool, dark place. Before adding it to the tank, heat it to the same temperature as the tank water; then

pour it gently into the outflow of the filter to enable it to blend in thoroughly. Never add salt directly to the aquarium.

The protein skimmer
The skimmer is responsible for removing organic matter from the water, which accumulates in the top section in the forms of foam and a yellowish liquid. When removing this waste material, take the opportunity to clean the skimmer's reaction chamber at the same time.

The fight against algae
This is just as complicated as in a freshwater tank. Some fish species, such as tangs, are predominately vegetarian and can help to prevent the proliferation of nuisance algae. Reducing the light intensity for a while can also help to restrict algal growth.

MAINTENANCE SCHEDULE FOR SALTWATER AQUARIUM

	Fish	Plants	Water	Substrate	Equipment
Daily	Feed them; observe their behavior and possible signs of illness; remove any dead fish.		Monitor the temperature.		Make sure that the filter (and/or pump) is working properly.
Twice a week			Monitor the pH and levels of ammonia and nitrites.		
Once a week			Monitor the specific gravity.		Clean the protein skimmer.
Every 2 weeks			Change 10 percent of the volume in the first month after the tank is set up.	Siphon off waste with the water change in the first month of operation.	Clean half the filter media, then the other half the following week. Clean the inside front glass and the cover glass/condensation tray.
Every month		If necessary, prune and tidy up any seaweed (macro-algae).		Siphon off waste at the same time as the water change	
Every 3 to 6 months					Service the air pump and filter motor, following the manufacturer's instructions.
Every 6 months					Change the fluorescent bulbs.
When necessary		Remove undesirable algae.	Readjust the water level.		Replace the filter media.

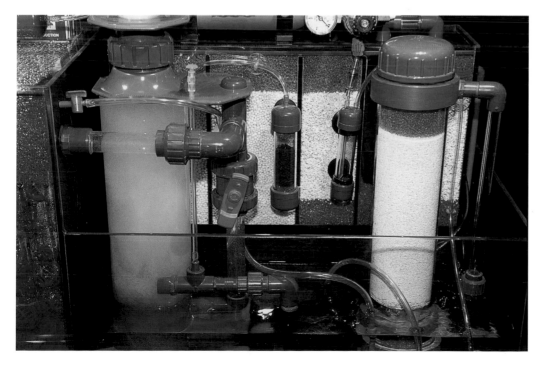

When an aquarium is contained within a cabinet, it is a good idea to assemble the water management system under the tank and ensure that it is readily accessible.

Introducing fish into a saltwater aquarium

Most of the advice offered in the chapter on introducing fish into a freshwater aquarium is equally applicable to saltwater. However, some specific characteristics must be taken into account for saltwater species.

Before buying a saltwater fish, first find out as much as you can about its diet and sociability.

A month-long wait

A saltwater aquarium demands even more patience than a freshwater one. A whole month is needed before the nitrogen cycle is established and the bacteria colonize the filter media. In most cases, introducing a fish into the tank during this interim period is the equivalent of a death sentence.

Schedule for the introduction of fish

As in freshwater, the smallest and most nervous species must be put in first. Larger or more lively fish must not be added for another two or three weeks, after a quarantine period in a separate tank.

Take care when choosing fish

Even more so than in a freshwater tank, you need to pay strict attention to the compatibility of the fish, whether they are of the same species or from a variety of species. Very few saltwater fish live in groups; in most cases, only one individual, or at most a couple, can be safely introduced.

Transportation

It is also necessary to take greater precautions in this respect than with freshwater fish. Only one saltwater fish can be transported in a plastic bag (along with a high proportion of air). However long the journey, the fish must be insulated from the cold. A Styrofoam box is an ideal insulated carrier.

The maximum size and number of fish

A saltwater tank cannot accommodate fish of the same size as a freshwater aquarium of similar dimensions. The larger marine species require significantly more space.

Tank dimensions (length x width x depth)	Tank volume	Max. adult length for tank size	Max. number
48 x 16 x 20 in. (120 x 40 x 50 cm)	63 gal. (240 L)	3–4 in. (8–10 cm)	4
60 x 20 x 20 in. (150 x 50 x 50 cm)	100 gal. (375 L)	6 in. (15 cm)	4
80 x 20 x 20 in. (200 x 50 x 50 cm)	132 gal. (500 L)	8–12 in. (20–30 cm)	3

Introducing the fish into the aquarium

Float the bag containing the fish on the surface of the tank to allow its temperature to adjust to that of the aquarium. This takes about 30 minutes.

Then open the bag and gently put the fish into the aquarium with a net. In order to avoid traumatizing the fish, it is advisable to transfer the fish to the tank when the lights are switched off.

How many fish in an aquarium?

Apply the same rule as in a freshwater tank, but allow 20 in.^{2} of surface area for $^1/_2$ inch of fish length. Accordingly, the table below shows the maximum number of fish for a range of saltwater tanks.

Tank dimensions (length x width)	Tank surface area	Max. total length at adult age for tank size	Max. no. of 4-in. (10 cm) fish	Max. no. of 6-in. (15 cm) fish	Max. no. of 8 to 12-in. (20–30 cm) fish
48 x 16 in. (120 x 40 cm)	768 in.^{2} (4,800 cm^2)	16 in. (40 cm)	4	*	*
60 x 20 in. (150 x 50 cm)	1,200 in.^{2} (7,500 cm^2)	24 in. (62 cm)	6	4	*
80 x 20 in. (200 x 50 cm)	1,600 in.^{2} (10,000 cm^2)	32 in. (83 cm)	8	5	3

*Fish of this size cannot be kept in captivity in this size of aquarium.

AQUARIUM
FISH AND PLANTS

HOW FISH LIVE

Every aquarist should have at least a passing knowledge of the anatomy and biology of fish. How do they breathe? What can they see? What do they eat? Why do some of them have barbels? The following section provides answers to such questions, allowing you to better understand the behavior of your fish and thereby adapt the conditions created in your aquarium more closely to their way of life. There is little point in keeping fish in a tank if you are not prepared to provide them with the best possible conditions. Each species has its own particular environmental needs, but there is one that is common to them all: healthy water conditions. In other words, filtration and regular partial water changes are the key to success in an aquarium. Another essential factor is appropriate feeding. It is important to remember that not all fish eat the same food. Many have a vegetarian diet, some are carnivorous and others are scavengers, but very few spurn portions of live food. Brine shrimp, bloodworms and water fleas (*Daphnia* spp.) are sources of fresh protein and vitamins that enhance the condition and coloring of fish, as well as encourage them to breed.

The external anatomy

In order to better appreciate the biology of aquarium fish, it is often helpful to know the name, location and function of the external features. Some of these body parts are exclusive to fish and play a highly specific role, particularly with respect to feeding and reproduction.

The coloring of male fish often becomes brighter during the mating season, as seen in this White Cloud Mountain minnow *(Tanichthys albonubes)*.

The skin

Fish are covered by a skin with overlapping scales that serves several roles, but predominantly a protective one. The scales are coated with mucus, forming a protective barrier against pathogenic organisms (i.e., those liable to cause disease). Some species, such as catfish, are even covered in bony plates.

Skin color changes in accordance with behavioral patterns, age and also sexual activity. These changes allow fish to pass messages to each other, while also providing fishkeepers with clues that provide a greater understanding of the fish's lifestyle. Furthermore, skin color serves as camouflage in some fish. The skin also plays a role in osmoregulation – the balance of fluids and salts within the body (see page 89).

The fins

The caudal fin propels a fish when it swims. The dorsal (there are sometimes two of these) and anal fins stabilize the fish and stop it from rolling, while the paired pectoral and pelvic fins enable a fish to brake and change direction.

Fin membranes are supported by rays, sometimes in the form of spikes that protrude from the fin, which may endanger other fish (or even the fishkeeper). This is the case with scorpionfish, whose fin rays are poisonous.

Characins (which include tetras) and catfish also possess another small fin between the dorsal and anal fins, called the adipose fin. It has no rays and does not appear to serve any purpose. Some fish species have modified fins. In gouramis, for example, the pelvic fins are reduced to a thin filament that has a tactile function. In male livebearers, the anal

The female White Cloud Mountain minnow generally sports more muted colors. This one is heavy with eggs.

The main external features

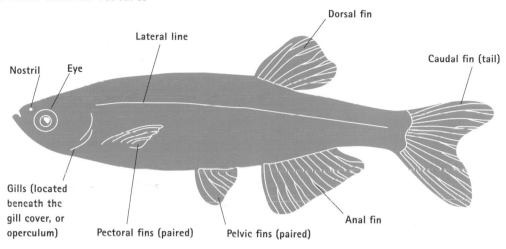

Nostril

Eye

Lateral line

Dorsal fin

Caudal fin (tail)

Gills (located beneath the gill cover, or operculum)

Pectoral fins (paired)

Pelvic fins (paired)

Anal fin

Below: Although the upside-down catfish (*Synodontis nigriventris*) mainly swims upside down, its fins are not unusual in any way.

fin is modified to form a reproductive organ called the gonopodium.

Some fish are specially bred with the aquarium hobby in mind, resulting in fish that have larger and more flamboyant fins, which provide a striking aquarium display

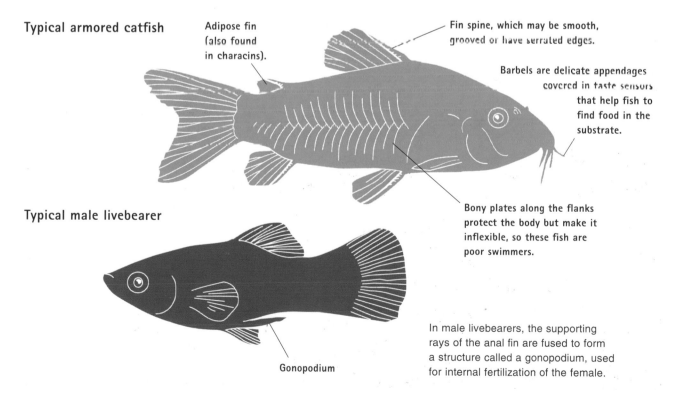

Typical armored catfish

Adipose fin (also found in characins).

Fin spine, which may be smooth, grooved or have serrated edges.

Barbels are delicate appendages covered in taste sensors that help fish to find food in the substrate.

Bony plates along the flanks protect the body but make it inflexible, so these fish are poor swimmers.

Typical male livebearer

Gonopodium

In male livebearers, the supporting rays of the anal fin are fused to form a structure called a gonopodium, used for internal fertilization of the female.

Above:
The barbels of the spotted pim *(Pimelodus pictus)* serve to detect food; they often also play a tactile role, for example, when they brush against obstacles.

The head
The eyes
The size of the eyes varies according to the fish's way of life. The eyes of fish that live in murky surroundings are small, but predators that live in clear water have larger eyes in order to better hunt their prey.

The positioning of the eyes on the side of the head endows fish with a wide field of vision. There are no eyelids.

The barbels
These are located near the mouth, on one of the two jaws (or occasionally on both). They are often found on fish that are bottom-feeders, as they help the fish detect food.

The operculum
This is a flap set behind the head to protect the gills; it plays a part in passing water over the gills for respiration.

The nostrils
These are not used for breathing, but are lined with sensory cells that detect odors. The sense of smell is highly developed in fish.

The mouth
The position of the mouth depends on the food a fish eats. In species that feed from the surface, the mouth points upward, whereas in those that feed off the bottom, it points downward. Most midwater fish have the mouth in a terminal position so they can approach their food head-on.

The lateral line system
Fish have no ears and cannot hear sounds, but they do perceive vibrations by means of the lateral line that runs along their flanks. A series of pores opens into a canal lined with pressure-sensitive cells that relay vibrations in the water to the brain.

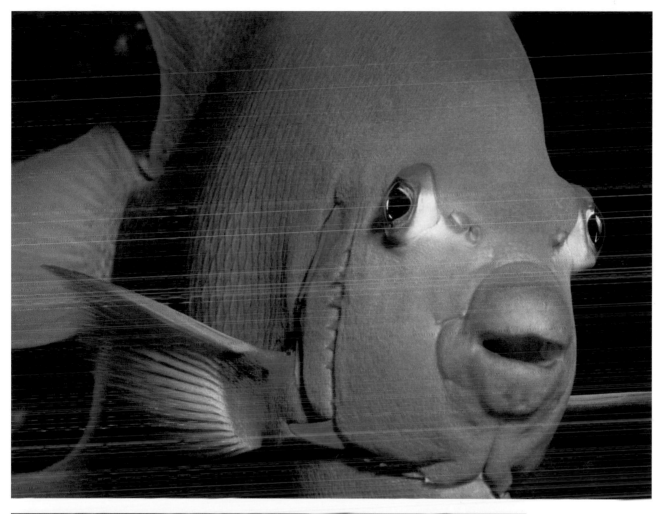

Above:
The eyes, which have
no lids, are set on the
sides of the head
close to the nostrils
on this queen angelfish
(*Holacanthus ciliarus*).

Left:
The lateral line, visible
as a series of dots
along the flanks of
these bloodfin tetras
(*Aphyocharax anisitsi*),
is a system specific
to fish that allows
them to feel pressure
changes in the water.

The organs and their roles

Some organs do not appear in this photograph, particularly those of the digestive and circulatory systems. The gut chemically breaks down food matter and allows food materials to be absorbed into the body. Food is mainly digested in the intestine. Digestive enzymes are produced in the pancreas and are then released into the intestine to break down food matter. Red and white blood cells, important for transporting oxygen and for the immune system, are produced and stored in the spleen. Old blood cells are broken down in the liver, producing bile, which is released as waste. The liver also stores glycogen, which is broken down into glucose as a source of energy.

Dorsal fin
This acts as a keel to prevent rolling. Some fish have two dorsal fins.

Adipose fin
This is only found in catfish and characins. It does not appear to play any specific role.

Caudal fin (tail)
Powerful muscles down either side of the body pull the tail, or caudal fin, from side to side to provide forward propulsion.

Lateral line
The lateral line system helps the fish to locate its surroundings in relation to itself. Using vibrations and electric signals, it provides a "sixth sense."

Kidneys
The kidneys act as a "body filter," removing wastes picked up and transported by the bloodstream. Kidneys also regulate the salt and water content of the fish's body and control the production of ammonia-containing urea.

Swim bladder
This gas-filled sac is used to regulate buoyancy in water. Not all fish have swim bladders; bottom-dwellers, such as many catfish, have no need to be buoyant, as they spend little time swimming in midwater.

Brain
The brain controls movement and bodily functions and also produces hormones. It controls memory and intelligence as well.

Mouth
This serves not only to catch prey, but also to inhale water, which passes out over the gills.

Eyes
Most fish have a wide field of vision, because the eyes are placed laterally on the head and can swivel independently. In species that live in bright clear water, the eyes are well developed and can detect color. They are smaller in fish that live in turbid conditions. Nocturnal fish that live in clear water have large eyes but see mainly in black and white.

Pectoral fins (paired)
These help a fish brake or change direction while it is swimming. They also stop the head from pitching up and down.

Gills
The thin membrane and large surface area of the gills allow oxygen from water taken in through the mouth to enter the bloodstream. Salts are absorbed through the gills by means of special chloride cells. The waste product, urea, is lost through the gills.

Pelvic fins (paired)
These help the fish to make small movements in the same way as the pectoral fins.

Anal fin
Together with the dorsal fin, the anal fin prevents the body from rolling sideways.

Heart
A simple, four-chambered heart pumps blood around the body, powering a circulatory system that carries oxygen, nutrients, waste and hormones.

The internal anatomy

Although hidden from view, fish have the usual bodily systems required to sustain life. The body cavity between the head and anal fin houses the digestive system, swim bladder, heart, kidneys and reproductive organs. Here we look at them in more detail.

Above:
The transparency of the glass catfish (*Kryptopterus bicirrhis*) leaves its skeleton open to view.

Right:
Frontosa *(Cyphotilapia frontosa)*, from Lake Tanganyika, is a good swimmer on account of its strong muscles.

The skeleton

Fish have a skull, a backbone and a series of smaller bones that we see as the "fish bones" on our plates. The backbone gives the body rigidity, while thinner bones support the fins. The skeleton contains calcium, which is absorbed from water or food.

The muscles

These are attached to the skeleton and are richly supplied with blood. To work efficiently, the muscles need a constant supply of nutrients and oxygen to enable fish to swim in the dense medium of water.

The nervous system

The nervous system is made up of the brain, which receives information from the environment through the sensory organs, and the nerves, which pass messages to the organs.

The bloodstream

The simple, four-chambered heart pumps blood to the gills and, newly charged with oxygen, this blood travels to the muscles and other body tissues. Oxygen-depleted, carbon-dioxide-rich blood is carried back to the heart for a new pumping cycle.

Gills

The gills are located on either side of the head behind the jaw. They are hidden from view beneath the gill covers (opercula). The four bony arches under each cover carry a double row of delicate, blood-rich gill filaments that appear bright red in healthy fish. The filaments are intricately folded and if spread out would almost equal the entire

surface area of the fish. Because of their thinness and the countercurrent flow of blood within them, the gills are highly efficient at extracting oxygen from and releasing carbon dioxide into the water that flows over them, on its way from the mouth and out through the gill covers.

Osmoregulation in a freshwater fish

The kidneys eliminate excess water from the body, but retain the mineral salts derived from the blood.

Water continually passes into the body through the skin and gills.

Chloride cells in the gills retain salts from the water flowing over them.

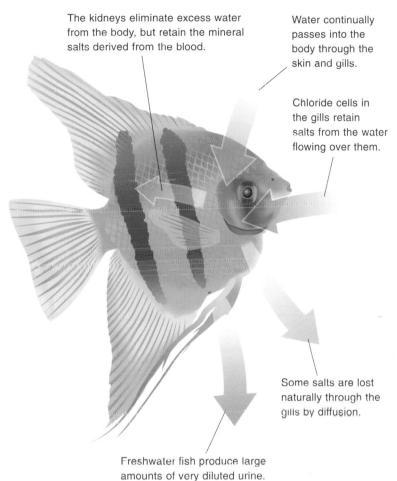

Some salts are lost naturally through the gills by diffusion.

Freshwater fish produce large amounts of very diluted urine.

Osmoregulation

To remain alive, it is vital for fish (and other living organisms) to maintain a constant balance of substances, such as salts and proteins, in their blood. At the heart of this balancing act is the battle to control the inflow and outflow of water and salts from the body, a process known as osmoregulation.

Freshwater fish contain a higher concentration of salts than the surrounding water. The natural tendency is for water to flow into the body by osmosis. Freshwater fish counteract this by producing a lot of diluted urine and retaining the salts they need.

Saltwater fish tissues are less "salty" than the sea and are thus in danger of dehydrating as water naturally passes out by osmosis. They cope by drinking seawater, eliminating salts through the gills and producing small amounts of concentrated urine.

The swim bladder

It acts as a buoyancy aid that enables the fish to retain its position in the water column. It is full of gases, mainly those found in air – oxygen, carbon dioxide and nitrogen – although the contents vary between types of fish. The swim bladder developed from the gut, and in some fish, the gut still has a connection with it. Bottom-dwelling fish have a reduced swim bladder or none at all.

The digestive system

Most fish have teeth in their jaws, which they use to grasp and shred their food. The "design" varies with the diet, from large, flesh-slicing incisors to masses of tiny "bumps" for rasping algae from rocks. Some fish, such as goldfish, have no teeth in their jaw, but use flat teeth at the back of the throat to grind their food

against a horny pad. Food passes into the digestive system and is broken down by enzymes into a fairly fluid form so that nutrients can pass into the bloodstream. Predatory meat-eating fish tend to have a more defined stomach but a relatively short digestive tract. Plant-eating fish have very long intestines in which colonies of bacteria break down much of the plant material in the diet. This process takes longer than digesting animal proteins.

The reproductive organs

The reproductive organs in fish are the testes (soft roe) in males and the ovaries (hard roe) in females. When the female is ready to spawn, the ovaries become swollen with eggs. The male fertilizes these with the milt (sperm) that travels through a duct to the genital aperture.

Behavior

You need to understand the habits and requirements of your fish in their natural setting if you are going to give them optimal living conditions. Particularly crucial are the relationships between species, or individuals of the same species, the concept of territory and the position of fish in the water column.

Plants and rocks allow fish to find their bearings and provide hideouts in case of danger.

Growth rates in fish

Unlike human beings, fish never stop growing, although the process slows down when they reach adulthood (which usually means sexual maturity). In an aquarium, the maximum size of a fish is often smaller than in the wild; this seems to depend on the volume of the tank. Fish that are small as adults grow more quickly than others.

Compatibility between species

Although most fish can live in a community, some species are less tolerant, or even aggressive, due to their territorial instincts: they defend the area where they live and chase off any intruders. Others only show aggression during the spawning period or when they are protecting their fry. Generally speaking, however, you should avoid putting fish species of very different sizes together, particularly if the bigger ones are carnivores.

Life in a group

Keeping groups of fish together is obviously only possible with peaceful species. It is not necessary, but if you choose to do so, it is highly recommended that you create a small group with at least five or six members. Allowing gregarious fish to live together puts them at ease and encourages them to reproduce.

In other species, only the young live in groups of small shoals; when they become adults they separate and live on their own. Their temperament is not necessarily affected by this transition, however, and most remain peaceful. However, it should be noted that many saltwater fish live alone and cannot tolerate any other fish in their territory.

Making the most of tank space

Some fish live at the surface and others prefer to stay near the bottom, but the majority swim in open water, so this area is usually the most heavily populated.

Pieces of decor often serve as reference points for fish, particularly if the fish are territorial. It is therefore important not to make any radical modifications to their surroundings once they have marked out their territories.

Environmental conditions

Fish live and breed in water that has specific characteristics, and these must be provided in the aquarium. For example, South American characins need soft, acidic water, while the livebearers of Central America and the cichlids of East African lakes prefer hard, alkaline water. Many species can tolerate different water qualities, but this does not necessarily mean that they will breed successfully in the home aquarium. Temperature is one of the most important parameters. Most tropical fish live in water

at 73–81°F (23–27°C), although some species will tolerate slightly cooler conditions.

Saltwater fish are a great deal more sensitive than freshwater species to variations in the water quality, especially with respect to temperature and specific gravity. It is therefore vital that these parameters are as stable as possible in the aquarium.

Regular water testing makes it possible to control the environment and prevent any fluctuations that may affect the livestock.

Corydoras catfish normally live on the bottom and rarely venture into open water.

Above: Fish such as this dwarf gourami *(Colisa lalia)* display their finest colors in water that matches their home environment.

Left: These cardinal tetras *(Paracheirodon axelrodi)* are typical characins that swim in open water. They like living in shoals of several fish.

Feeding

Food must supply proteins, fats, carbohydrates, vitamins and minerals – with protein and fat being a fish's primary source of energy. A high-quality diet should include all these elements in the correct proportions, according to the specific nutritional needs of the various species. Prepared and live foods from aquarium stores have made feeding livestock safe and easy.

Herbivorous fish

Truly herbivorous fish are fairly rare. Two examples are freshwater catfish in the Loricariidae family and saltwater tangs. There are other species that are largely herbivorous, but they will also accept animal food.

Omnivorous fish

The majority of fish are omnivores: they are opportunists that feed on both animal and vegetable foods. Omnivorous fish are undoubtedly the easiest species to keep.

Carnivorous fish

Carnivorous fish are predators that feed on live food, particularly insects, worms, crustaceans and even other, smaller fish. There are some species, such as saltwater angelfish, that eat sponges or corals.

Dried food

Dried food is the most popular aquarium food – and the easiest to use. It is formulated to provide a balanced diet, although this does not preclude the need for supplements. Dried foods are available in different forms, including floating flakes, granules of various densities, wafers, tablets and pellets, all in a range of sizes. There are even fine powdered foods suitable for fry.

All fish need to eat

Ensure all the fish in your tank have the opportunity to eat. Sometimes, more enterprising fish will grab more than their fair share. Feed these fish in one corner of the aquarium, then offer food to more timid fish in the opposite corner. When offering flakes, hold a pinch under the water for a moment before releasing it so that some will sink, otherwise surface-feeders will take most of the floating food. Bottom-dwellers should be fed last of all, as the other fish, already sated, will ignore food on the substrate that is intended for species that will not swim upward to feed.

Opposite page:
Some species of saltwater fish are virtually herbivorous and graze on algae. Others are decidedly carnivorous and appreciate live prey in the aquarium.

Flakes contain all the nutrients necessary for fish health.

Flakes based on brine shrimp, ideally enriched with vitamins, are appreciated by some saltwater fish.

As an alternative to flakes, dried food is available in the form of granules of various consistencies.

Floating pellets are ideal for large fish (that have correspondingly big mouths).

Sinking granules drop down to the bottom-feeders.

Tablet foods can be stuck onto the aquarium glass for midwater-feeders or left to fall to the substrate for bottom-feeders.

Live prey is much appreciated by most fish, particularly saltwater species reluctant to eat prepared foods. It is sometimes the only way to feed these fish. Here, a decorated dartfish *(Nemateleotris decora)* "captures" a fragment of shrimp.

Freeze-dried foods

Various small larvae, shrimp and other invertebrates are available in freeze-dried form as aquarium food. Fish relish them as a change from prepared foods.

Fish flakes are the most common form of dry food.

Frozen food

A wide range of frozen foods are available, including worms and crustaceans such as brine shrimp, river or saltwater shrimp, and water fleas (*Daphnia* spp.). Frozen plant-based food is also available.

True and false worms

Tubifex worms are small and thin and, because they live naturally in polluted mud, are best avoided as live food. They are perfectly safe in freeze-dried form, however. Bloodworms are not true worms but rather the larvae of a non-biting midge. They are available in frozen form, as well as in plastic bags, partly filled with water, from aquarium stores and bait shops. When offering bloodworms, do not simply tip the bag into the aquarium. Strain the worms through a fine-meshed nylon net and empty the net under the surface. (Keep the net solely for this purpose.) You can also put worms into a floating feeder that contains the food in one location at the surface. Both these "worm" foods are rich in protein and provide welcome variety in the diets of aquarium fish.

Live foods

These are undoubtedly the favorite food, not only of carnivorous fish, but also of omnivores. They are sometimes indispensable for species that refuse dried food in the aquarium, particularly in the case of saltwater fish. The most widely used live foods in the hobby are water fleas (*Daphnia* spp.), bloodworms and adult brine shrimp.

Brine shrimp

Brine shrimp are primitive crustaceans that reach a length of about $^1/_2$ inch (1.25 cm) as adults. They are widely available as a live food in small plastic bags partially filled with water.

Brine shrimp live in salty marshes that dry up in summer. In order for the species to survive, they lay eggs known as cysts that resist both heat and dehydration. They can be kept for several months or even years (although in this case the hatching will be less successful).

The eggs hatch 24 to 36 hours after the cysts are put into saltwater. The larvae (or nauplii) that emerge are an excellent fry food. The techniques for producing nauplii are described in the chapter on breeding. Adult brine shrimp also make good food for mature fish.

Left: Bloodworms, rich in protein, attract fish by their movements.

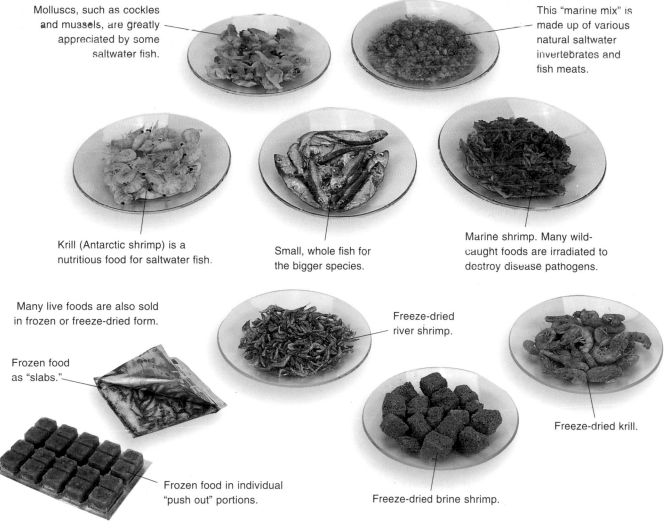

Molluscs, such as cockles and mussels, are greatly appreciated by some saltwater fish.

This "marine mix" is made up of various natural saltwater invertebrates and fish meats.

Krill (Antarctic shrimp) is a nutritious food for saltwater fish.

Small, whole fish for the bigger species.

Marine shrimp. Many wild-caught foods are irradiated to destroy disease pathogens.

Many live foods are also sold in frozen or freeze-dried form.

Freeze-dried river shrimp.

Frozen food as "slabs."

Freeze-dried krill.

Frozen food in individual "push out" portions.

Freeze-dried brine shrimp.

Above:
Aquarists can prepare their own mixtures based on fish and meaty foods.

Right:
Liquid food containing suspended food particles are sold for young fry.

Fresh food

Fish can also be given a wide array of items from their owners' diet. Carnivores can be fed mussels, cockles, shrimp, fish and even meat. In the latter case, it is preferable to use white meat such as chicken, as the fat content is lower. Obviously, all these foods must be cooked before they are served to aquarium fish.

As for vegetables, spinach, lettuce and small peas are all appealing to herbivorous fish, but must first be briefly boiled in water and cooled before being fed to them. It is also possible to make your own pâté from a mixture of the above ingredients in varying proportions: predominantly vegetables for herbivores, a mixture of equal parts of meat and vegetables for omnivores and meat alone for carnivores.

Food distribution

It is essential to respect the different feeding habits of your aquarium inhabitants. Some eat by day, others at night (especially bottom-dwellers). It is therefore practical to distribute food after "lights out," making sure that the food is eaten within 15 minutes by the open-water species.

Food that drops to the substrate for the bottom-dwelling fish should all be gone by the following morning. It will take a few days of adjustment before you are able

Feeding saltwater fish

In the case of marine fish in particular, it is important to provide a varied diet and to alternate the foods that are offered. This is often more complicated than feeding freshwater fish. Prepared food plays a less important role, and preference must be given to live prey or frozen food for carnivorous fish (which also enjoy mussels, fish and meat). Some fishkeepers prepare a mixture with these ingredients, binding them with gelatin and freezing the "block" for long-term use. Herbivorous species can graze on the algae that grow on the decor and tank glass, but they must also be supplied with vegetables – cooked spinach is a good option. Prepared seaweed is available in sheet form, which can be clipped onto the glass.

Above:
This battery-operated auto-feeder is easy to program to release dried food at regular intervals. This is a practical strategy for vacation feeding.

to calculate the amount of food your fish require. It is important to avoid overfeeding, as surplus food will decompose (thereby consuming oxygen) and pollute the water, with a subsequent risk to the health of the fish and other livestock.

Alternating a variety of foodstuffs – even different brands of dried food – is advisable for fish that accept a broad diet.

Above:
Lettuce is appreciated by saltwater species with a plant-based diet like this yellow tang (*Zebrasoma flavescens*). Boil it quickly before serving. Spinach is also suitable.

Left:
The mouth of this sailfin molly (*Poecilia latipinna*) is adapted to surface feeding.

Healthcare in the aquarium

It is all too easy to underestimate the importance of healthcare in your tank. In an enclosed setting like an aquarium, diseases can break out and spread very quickly, and they are often contagious. Some are common and can be cured easily with off-the-shelf remedies, while others are rare and sometimes incurable. Diseases have a variety of origins: pathogenic organisms and parasites, poor water conditions, low-quality food, overfeeding and stress.

Prevention is better than a cure

This proverb is also applicable to fish. In order to avoid disease, it is important to provide them with healthy surroundings. This means providing the best possible water conditions for the species you keep. Effective filtration and aeration, and therefore oxygenation, are very important factors for healthy fish, although appropriate feeding should not be neglected. A balanced and varied diet is one of the best means of avoiding disease. A badly nourished fish becomes weak and is more likely to become sick. Stress can also lead to ill health, whether caused by territorial incompatibility or disturbance from outside the tank.

Some symptoms are clearly apparent. Here, a small barb has a rotting caudal fin.

Health problems in fish

Infectious diseases are caused by pathogens such as viruses, bacteria, fungi and parasites. They are often highly contagious.

Behavioral problems are very often caused by low-quality water, as well as certain parasites. Weight loss or physical deformities are often due to a poor diet, but these symptoms may also be caused by one of the diseases mentioned.

Fungus can develop after a wound is sustained and can quickly spread over a large part of the body.

Viral infections

Viral infections are less common than might be imagined and often prove difficult to detect. Probably the most common is lymphocystis, which produces white swellings resembling tumors on the fins and skin. There is no treatment, but the growths may disappear on their own.

Bacterial infections

These are fairly common, often appearing after a fish is wounded. Bacterial infections produce ulcerations and reddened areas of skin. Some attack the nervous system, giving rise to erratic swimming (shimmying). These infections can be treated with antibiotics.

Fungal infections

These infections are mainly caused by a fungus, *Saprolegnia*, which grows on wounds or invades the body at the site of a missing scale. Fungal growths can also appear on newly laid eggs in a breeding tank. An infection is signaled by the appearance of small, cottony white clumps on the body or fins. It is easy to treat with medications for sale in aquarium stores, but you must act promptly as the fungus can spread over the body and go on to attack other fish. Preventative treatment is available specifically for eggs. In all cases, it is vital to follow the dosages recommended in the manufacturer's instructions.

The erosion of muscles on the head and lateral line of this tang is probably caused by a lack of vitamin A, rather than the action of microorganisms.

Right: The fin at the top is in good condition with smooth margins. On the bottom one, fit rot has set in, and the degeneration of the membrane between the fin rays is clearly apparent.

Parasitic infections

These are triggered by protozoans (microscopic single-celled animals), worms or crustaceans. The most common infection caused by a protozoan parasite is known as "ich" or "whitespot."

Worms called flukes can infest the skin and gill filaments, as well as some internal organs. Roundworms can grow within the gut. Crustacean parasites include fish lice and anchor worms. These attach to a fish's skin and, as with worm parasites, can be difficult to eradicate.

Diseases with environmental causes

Life-threatening respiratory diseases are brought on by a lack of oxygen or by the presence of toxins, such as ammonia and nitrites, in the water. Water changes and increased oxygenation are essential treatments in these cases. Poor environmental conditions can also lead to bacterial infections that cause the degeneration of

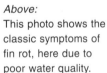

Above:
This photo shows the classic symptoms of fin rot, here due to poor water quality.

Right:
It is vital to treat fin rot promptly, here on the dorsal fin of a dwarf gourami *(Colisa lalia)*, to prevent the decay from spreading and killing the fish.

the fin membranes, a condition known as fin rot. This problem often disappears when the fish is given better quality water. In the most serious cases, treatment with suitable antibacterial medications may be necessary.

Right: Ich has spread all over the body of this harlequin rasbora (*Rasbora heteromorpha*). It is in urgent need of treatment.

Ich

Both freshwater and saltwater fish can suffer from ich, also called whitespot, in which pinhead-sized spots appear on the body and fins. In freshwater fish, they are caused by the protozoan parasite *Ichthyophthirius multifiliis*, which lives under fish skin and matures into cysts that "punch" out and fall to the bottom of the aquarium. Each cyst can produce at least a thousand free-swimming parasites that must find a fish host within 24 hours in order to survive. It is at this free-swimming stage of the cycle that the parasite is most susceptible to treatment. Although it is advisable to isolate sick fish, the tank as a whole must also be treated.

The parasite that causes an equivalent condition in saltwater fish is *Cryptocaryon irritans*. In both cases, effective treatments are readily available.

This saltwater fish, a powder blue tang (*Acanthurus leucosternon*), is afflicted by ich, caused by *Cryptocaryon irritans*.

A treatment/quarantine tank

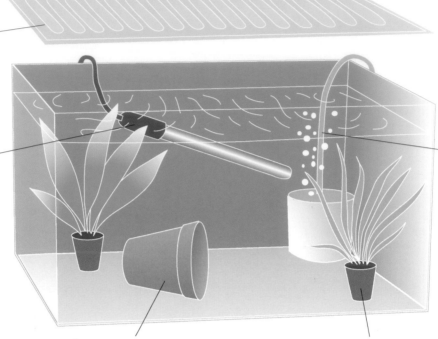

Cover the tank to prevent stressed fish from jumping out.

Position the heater-stat well above the bottom of the tank so that fish are not tempted to hide beneath it, where they may suffer burns. Ideally, install a heater guard.

A small internal filter ensures that the water remains clean. Do not include any activated carbon, as it will remove medications from the water.

A flowerpot (thoroughly washed and scrubbed) can provide welcome shelter as a "cave."

Plastic plants can provide shelter to fish stressed by their transfer to the treatment/quarantine tank.

The treatment aquarium

This is the best way to treat ailing fish. A treatment aquarium does not have to be large – a few gallons will do. Lighting is unnecessary – natural light will not disturb the fish, which are already delicate on account of their sickness. The water must be heated and filtered through a small internal filter containing a block of foam. You can provide shelters (a few rocks, for example, or a flowerpot) that have been cleaned in very hot water beforehand. Plastic plants will provide a reassuring sanctuary for stressed fish. A substrate serves no purpose and is not required. The treatment aquarium can also be used as a quarantine tank to isolate new fish before they are transferred to the display aquarium.

Right: Calculate the correct amount of medication to be added to the tank, following the manufacturer's directions carefully. Fill a bottle or plastic jug with water from the aquarium and add the medication to it.

Treatment

It is advisable to use a quarantine tank to isolate the most severely affected fish, but be sure to also treat the aquarium itself.

Aquarium retailers sell medications specifically formulated to treat the most common diseases. Many of these medications are supplied in a liquid form and are very easy to use.

Euthanasia

For incurably sick fish, the only way to relieve their suffering is to dispose of them humanely. One method is to obtain the fish anesthetic MS222 from a veterinarian and leave the fish in a solution of this for several hours. Never put live fish in the freezer or flush them down the toilet. If you are unsure about how to deal with this situation, seek help from your aquarium dealer or fellow hobbyists. You must take action to safeguard your other fish.

Any medications must be correctly administered, and it is vital to pay strict attention to the recommended dosage and treatment schedule.

Some fishkeepers, with the best of intentions, increase the doses, put different medications in the tank every day or even mix different medications together. This is not only ineffectual, but also dangerous, as the medications have been thoroughly tested to work in the way recommended on the packaging and not in combination or at higher doses.

Before adding medications to the aquarium, first dilute them in a plastic bottle or jug of water from the tank. In some cases, medications can impair the functioning of a biological filter for some time. To compensate for this, increase the oxygenation of the water.

Top far left:
Increase oxygenation when treating sick fish, especially if the biological filter has to be switched off.

Left:
Gently introduce the water that contains the diluted medication back into the aquarium. Keep any utensils solely for aquarium use.

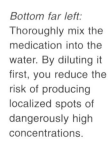

Bottom far left:
Thoroughly mix the medication into the water. By diluting it first, you reduce the risk of producing localized spots of dangerously high concentrations.

In emergencies, slightly increasing the water temperature can limit the impact of a disease without affecting the plants and fish.

If fish are treated in a quarantine tank, they must not be put back into their original aquarium until they are completely cured.

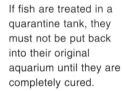

Emergency solutions

If no medications are in hand, you can resort to some emergency strategies that are intended to curb the development of diseases rather than necessarily cure them. First of all, intensify the aeration. Next, carry out a partial water change, avoiding any abrupt change in water temperature. Finally, increase the temperature slightly – some pathogens will not tolerate the extra heat.

Using salt as a treatment

Salt (either kosher, pickling or aquarium salt) can prove effective against fungus and fin rot. Beware, however, some fish that live in freshwater will not withstand excessive levels of salt.

Use a level teaspoon (5 ml) of salt per quart (1 L) of water to be treated, or two teaspoons (10 ml) in the case of species that live in hard water. The salt must be diluted beforehand in a jug, then poured into the aquarium water, a little at a time over several days.

Once the fish are cured, bring the water back to its original state by changing 10–20 percent of the water every day.

After recovery

When fish are in poor health, they grow weaker and eat less than usual. Once they are cured, they resume their usual activities. Be sure to provide them with a varied, high-energy diet that allows them to recover their full vitality and resume their normal behavior. Only then can they be removed from the treatment tank to become part of the main aquarium display once again.

MAIN AILMENTS AND DISEASES

Symptoms	Possible causes	Treatment
Loss of balance, jerky swimming (abrupt start followed by halt), fast breathing.	Excessively high levels of ammonia or nitrites. Lack of oxygen. These symptoms can also indicate a bacterial infection.	Make an immediate 25 percent water change and then partial water changes for a few days. Increase oxygenation. Treat any bacterial infection with a medication.
Gulping air at the surface, rapid breathing.	Lack of oxygen. Ammonia or nitrite poisoning.	Increase oxygenation. Improve water quality.
Erratic swimming (shimmying).	Bacteria attacking the nervous system.	Proprietary (antibiotic) medication.
Rubbing against decor (flashing), excessive mucus, red skin.	External parasite.	Proprietary medication.
Weight loss, retarded growth.	Internal parasite (often a worm).	Proprietary (worming) medication.
Swollen abdomen, wounds on fins.	Dropsy caused by bacteria.	Proprietary (antibiotic) medication.
Protruding operculum, unsteady swimming (freshwater fish).	Parasitic gill fluke on the gill filaments.	Proprietary medication. Salt.
Skin ulcers, reddened fins/skin.	Bacterial infection.	Proprietary (antibiotic) medication.
Cottonlike layer (rare in saltwater fish).	Fungal infection caused by *Saprolegnia*.	Proprietary medication. Salt.
Small white spots (freshwater fish).	Ich parasite (*Ichthyophthirius*).	Proprietary medication. Increase temperature.
Small white spots (saltwater fish).	Saltwater ich (*Cryptocaryon*).	Proprietary medication. Increase temperature.
Dusting of fine white or gold spots.	Velvet (*Oodinium*) in freshwater fish. Coral fish disease (*Amyloodinium*).	Proprietary medication. Remove fish and increase temperature to speed life cycle.
Swollen eyes (pop-eye).	Bacterial or sometimes viral infection.	Proprietary medication.
Fin rot.	Bacterial infection.	Proprietary medication. Salt.
Degeneration of the barbels (freshwater fish).	Poor water quality. Unsuitable substrate material.	Partial water changes. Replace the substrate.
Hole in the fish's head. (Head-and-lateral-line disease).	Many possible causes: poor water quality, lack of vitamins, protozoan parasites, bacterial infection, poor environment, viral infection.	Partial water changes. Supplementary vitamins. Diet based on live and fresh foods. Try antibiotics.

Once fish have been cured, a rich and varied diet will allow them to build up their strength and quickly get back to normal.

FRESHWATER FISH

Of the 15,000 to 20,000 species of fish living in the world's freshwater, only about 1,500 are suitable for keeping in an aquarium. Many are unsuitable because of their lack of visual appeal, others for their enormous size or undesirable behavior! And yet others are under import or other restrictions, so do research any unusual specimens you wish to acquire.

When the time comes for you to decide what fish best suit your particular requirements, there is a remarkably wide choice, particularly from the many species that are selectively bred in different varieties (above all, the livebearers).

The Amazon Basin is by far the greatest supplier of aquarium fish. From the magnificent cardinal tetras (Paracheirodon axelrodi) to the superb (albeit uncouth) piranhas, the range is huge, not only in color and size but also in behavior. Africa offers fish with widely varying characteristics, depending on whether they live in the Great Lakes of the Rift Valley (Tanganyika, Malawi and Victoria) or in rivers. As for Asia, the diversity is similarly exceptional, with gouramis and related species living in the overheated, oxygen-starved waters of paddy fields, and others adapted to thrive in the slightly salty water found in estuaries. It is this extraordinary biodiversity that forms one of the major attractions of keeping an aquarium.

Labyrinth fish (Anabantids)

In the wild, these fish live in stagnant pools that are low in oxygen. To overcome this, they have developed a special organ above the gills – the labyrinth organ – that enables them to extract vital oxygen from air they gulp air at the surface. They are strong fish that build a bubblenest at breeding time; this protects both the eggs and the fry. They are not fussy eaters and willingly accept prepared food.

BETTA
Betta splendens
Size: 2–3in. (5–7.5 cm)

Two males put together in the same aquarium will automatically fight – hence they are also known as Siamese fighting fish.

Origin:	Number of fish per aquarium:	Diet:
Thailand, Cambodia	Only 1 male, but can be kept with 1 or 2 females	Dried food, plus small live or frozen food

Betta

Several varieties of betta, or Siamese fighting fish, are bred for aquariums. They are distinguished by their bright coloring and large fins. If the fins are damaged fungal infections can set in. For the fins to remain intact, it is a good idea to raise bettas with peaceful species, including those of their own family.

The females (much more lackluster and with smaller fins) are tolerated by the males, but males will fight among themselves if they share the same space. If you put a mirror in front of the aquarium, a betta will even attack its own reflection! This behavior has been exploited in their home region, with public

Other fighting fish

The red betta, *Betta coccina* (2–3 inches/5–7.5 cm), a timid fish that requires live prey, is rarely found in the aquarium hobby. Like the other fighting fish, it lays its

eggs in a bubblenest. This is not true of the following. The one-spot betta, *Betta unimaculata*, is bigger (4–5 inches/10–12.5 cm), although the female is smaller and has drab colors. It does not make a bubblenest, but lays eggs on the substrate. The male incubates the eggs in his mouth for about ten days. During this period, he often remains hidden in the vegetation. As soon as the fry leave the shelter of his mouth, they start to look for food; they will accept brine shrimp nauplii. The Javan mouthbrooder, *Betta picta*, is one of the smallest fish in this group, rarely exceeding 2 inches (5 cm). These three fish are more peaceful than *Betta splendens*, but with less developed fins.

Left: Betta unimaculata

fights staged between males, complete with betting.

These fish like neutral water that is not too hard and is fairly warm – a sharp drop in temperature can trigger illness.

When keeping bettas in the aquarium, it is important that they have easy access to the surface, both to take gulps of air and to build a bubblenest during the breeding period. If these conditions are satisfied, bettas are not difficult to keep.

Peaceful (crescent) betta

The peaceful betta is slightly smaller than the more familiar betta and has less prominent fins. It reproduces in the same way (i.e., with a bubblenest), but the process is simpler. The female (less colorful than the male, and with shorter fins) can lay more than 100 eggs, which hatch in two days. The fry will readily feed on newly hatched brine shrimp (nauplii).

PEACEFUL (CRESCENT) BETTA
Betta imbellis
Size: 2 1/2 in. (6.5 cm)

The common name is justified. When two males cross paths, they put on a display by opening their fins, but they do not attack each other.

Origin:	Number of fish per aquarium:	Diet:
Malaysia, Thailand	1 pair	Dried food, small live or frozen food

The labyrinth fish aquarium

1 GIANT GOURAMI
Osphronemus goramy

Family: Osphronemidae
Size: up to 24 in. (60 cm)

Origin:	Number of fish per aquarium:	Diet:
Asia	1 per 80–135 gal. (300–510 L) tank	Omnivorous

2 CHOCOLATE GOURAMI
Sphaerichthys osphromenoides

Family: Belontiidae
Size: 2 in. (5 cm)

Origin:	Number of fish per aquarium:	Diet:
Malaysia, Indonesia	1 pair per 15-gal. (55 L) tank	Live food, flakes

3 THICK-LIPPED GOURAMI
Colisa labiosa

Family: Belontiidae
Size: 3–4 in. (7.5–10 cm)

Origin:	Number of fish per aquarium:	Diet:
Burma, India	1 pair per 20-gal. (75 L) tank	Live food; omnivorous

4 HONEY GOURAMI
Colisa chuna

Family: Belontiidae
Size: 2 in. (5 cm)

Origin:	Number of fish per aquarium:	Diet:
India, Bangladesh	1 pair per 15 gal. (55 L)	Live food, flakes

5 PYGMY GOURAMI
Trichopsis pumila

Family: Belontiidae
Size: 1–2 in. (2.5–5 cm)

Origin:	Number of fish per aquarium:	Diet:
Thailand, Vietnam, Sumatra	2 pairs per 15-gal. (55 L) tank	Small live food, flakes

6 CROAKING GOURAMI
Trichopsis vittata

Family: Osphronemidae
Size: 2–3 in. (5–7.5 cm)

Origin:	Number of fish per aquarium:	Diet:
Southeast Asia	2 pairs per 25-gal. (95 L) tank	Live food; omnivorous

Pearl gourami

This tranquil fish adapts easily to a community tank; it is undemanding in regards to water quality and is suitable for novices. A pair will build a bubblenest for breeding. Its long, filamentous pectoral fins play a tactile role and, like the other fins, can attract the attention of more belligerent fish. It is therefore a good idea to raise pearl gourami with other peaceful species.

Dwarf gourami

This fish is ideally kept in an aquarium with other peaceful species. It appreciates regular partial water changes, as poor-quality water can damage its fins. It reproduces easily, laying its eggs in a bubblenest.

Kissing gourami

The name is derived from its unusual behavior: two fish often seem to be kissing each other on the mouth. This behavior has no sexual connotations, but serves to establish a fish's social position in a group or territory. This is a very peaceful fish that likes to eat small algae (making it a useful asset in an aquarium). It is very difficult to distinguish a male from a female. This gourami does not build a bubblenest.

1 DWARF GOURAMI
Colisa lalia
Size: **2 in. (5 cm)**
The male has blue and red stripes, while the female is a silvery gray.

Origin:	Number of fish per aquarium:	Diet:
India	1 pair	Small live and frozen food, dried food

2 KISSING GOURAMI
Helostoma temminckii
Size: **6 in. (15 cm)**
This fish belongs to the Helostomatidae family, closely related to the other gouramis. It is normally a pinkish color, but there is also a less common, green variety.

Origin:	Number of fish per aquarium:	Diet:
Thailand, Java	2 to 5	Dried food based on plants, small live and frozen foods

Blue gourami

The males are distinguished by their more pointed dorsal fin. Although older males can be aggressive toward each other, they are generally peaceful in a mixed community aquarium. They lay their eggs in a bubblenest and produce large broods.

3 PEARL GOURAMI
Trichogaster leeri
Size: **6 in. (15 cm)**

The male has brighter colors and longer fins than the female.

Origin:	Number of fish per aquarium:	Diet:
Thailand, Malaysia, Sumatra, Borneo	1 pair	Dried food, small live and frozen foods

4 BLUE GOURAMI
Trichogaster trichopterus
Size: **4 in. (10 cm)**

The blue body can be set off by areas of black, in the form of patches or two round spots.

Origin:	Number of fish per aquarium:	Diet:
Burma, Thailand, Malaysia, Indonesia	1 pair	Dried food, small live and frozen foods

The Asian aquarium

1 BANDULA BARB
Puntius bandula

Family: Cyprinidae
Size: 1¹/₂–2¹/₂ in. (4–6.5 cm)

Origin:	Number of fish per aquarium:	Diet:
Asia	Up to 10 per 25-gal. (95 L) tank	Live and frozen foods, flakes

2 GREATER SCISSORTAIL
Rasbora caudimaculata

Family: Cyprinidae
Size: 4–5 in. (10–12.5 cm)

Origin:	Number of fish per aquarium:	Diet:
Indonesia, Malaysia, Thailand	5 per 25-gal. (95 L) tank	Live foods, vegetable-based flakes

3 BROWN'S RED DWARF FIGHTER
Betta brownorum Matang

Family: Belontiidae
Size: 1¹/₂ in. (4 cm)

Origin:	Number of fish per aquarium:	Diet:
Borneo	1 pair in a species aquarium	Live and frozen foods, flakes

4 TIGER BOTIA
Botia helodes

Family: Cobitidae
Size: 6–8 in. (15–20 cm)

Origin:	Number of fish per aquarium:	Diet:
Thailand, Cambodia, Laos	1 per 25-gal. (95 L) tank	Omnivorous

5 SKUNK BOTIA
Botia morleti

Family: Cobitidae
Size: 3–4 in. (7.5–10 cm)

Origin:	Number of fish per aquarium:	Diet:
Thailand	2 or 3 per 25-gal. (95 L) tank	Omnivorous

6 PEARLY RASBORA
Rasbora vaterifloris

Family: Cyprinidae
Size: 1¹/₂ in. (4 cm)

Origin:	Number of fish per aquarium:	Diet:
South India, Sri Lanka	1 shoal of 10 fish per 25-gal. (95 L) tank	Flakes, various small live foods

Characins

Most characins originate in South America. They are recognizable by a small supplementary fin – the adipose fin. They like to live in groups, in soft, acidic water, and they are fairly peaceful. In the wild, characins feed mainly on small insects and their larvae. Another related family, the Alestidae, is discussed at the end of this chapter.

1 CARDINAL TETRA
Paracheirodon axelrodi
Size: **2 in. (5 cm)**

Sometimes confused with the neon tetra, the cardinal tetra is distinguished by the bright red coloring over all the lower part of its body.

Origin:	Number of fish per aquarium:	Diet:
Parts of Colombia and Brazil	Minimum of 6	Small live and frozen foods, dried food

Cardinal tetra

The vivid coloring of the cardinal tetra makes it one of the most popular aquarium fish, but it is far from easy to breed. It requires acidic water (pH 6 to 6.5) with negligible hardness and a temperature of 77°F to 81°F (25°C to 27°C). This fish does not do well in a tank that has recently been set up – it is safer to put cardinal tetras into an established tank. Dense planting provides it with shelter if it becomes unsettled.

It is peaceful and lives in shoals, while also accepting the presence of species of the same size, such as those from its own family, all native to the same region of South America.

To breed these fish, put a pair into a small tank, with peat to acidify the water. The female lays up to 500 eggs, at night or at daybreak; incubation lasts for two days. The fry like very subdued lighting. When they grow bigger, they must be gradually acclimatized to the water in the aquarium that will be their final home.

Black neon tetra

The black neon tetra prefers acidic or neutral water that is fairly soft. Males are more slender than females, which lay their eggs on fine-leaved plants. Eggs hatch 36 hours later and the fry feed on brine shrimp nauplii. They grow quite quickly and will, in turn, be able to breed at the age of 8–10 months.

Neon tetra

The highly popular neon tetra relishes an open area in a planted aquarium so that it can swim around in a shoal. Its colors can be set off by a dark substrate. Breeding takes place in a small tank with acidic water (pH 6 to 6.5) of limited hardness, under subdued lighting. The eggs are laid on fine-leaved plants and hatch in 24 hours. The female can lay up to 300 eggs.

3 BLACK NEON TETRA
Hyphessobrycon herbertaxelrodi
Size: 1¹/₂ in. (4 cm)
This fish is peaceful and likes to live in a small shoal.

Origin:	Number of fish per aquarium:	Diet:
Brazil	Minimum of 4 to 6	Live and frozen foods, dried food

4 GLOWLIGHT TETRA
Hemigrammus erythrozonus
Size: 1¹/₂ in. (4 cm)
A group of glowlight tetras creates a stunning visual effect in the display aquarium. This fish is ideally suited to novice aquarists.

Origin:	Number of fish per aquarium:	Diet:
Guyana	Minimum of 4 to 6	Dried food, small live and frozen foods

2 NEON TETRA
Paracheirodon innesi
Size: 1¹/₂ in. (4 cm)
This fish owes its name to its bright colors. The neon tetra is prolific, but the fry may prove hard to raise successfully.

Origin:	Number of fish per aquarium:	Diet:
Peru	Minimum of 8 to 10	Omnivorous, but appreciates small live and frozen foods

Glowlight tetra

The glowlight tetra generally proves to be hardy in an aquarium. Its coloring is enhanced by a healthy diet and high-quality water that is slightly acidic and soft. Mature females – less pointed than the males – lay their eggs on plants.

Emperor tetra

This is a calm fish that likes swimming but is also inclined to take refuge in the shadows among vegetation. A densely planted tank with shade is therefore essential, along with an area of open water. It prefers neutral, not overly soft water, particularly for breeding. Males are more brightly colored than females.

1 EMPEROR TETRA
Nematobrycon palmeri
Size: **2 in. (5 cm)**

This is ideal for a characin aquarium, although it is difficult to breed (and not a prolific egg-layer).

Origin:	Number of fish per aquarium:	Diet:
Northern part of South America, particularly Colombia	Minimum of 4 to 6	Dried food, but prefers small live and frozen foods

Bleeding heart tetra

This fish enjoys the company of peaceful species of the same size for swimming in open water. It has a low nitrate tolerance, so regular partial water changes are essential. The female is less colorful than the male, and its fins are shorter. The anal and dorsal fins are less prominent in younger fish. Breeding is fairly difficult; it requires a small tank with moderate lighting and peat (to obtain a pH of 6 to 6.5). The female lays 20 to 30 eggs, which hatch two or three days later. The fry start to search for food the day after they hatch.

Rosy tetra

This fish is distinguished from a closely related species, serpae tetra *(Hyphessobrycon eques)*, by the absence of a black spot behind its head. It can be kept in a group with other characins. It lays its eggs on the bottom – at which point the parents must be removed, as they are likely to eat the eggs.

2 BLEEDING HEART TETRA
Hyphessobrycon erythrostigma
Size: **2¹/₂ in. (6.5 cm)**

This fish owes its name to the red spot behind the operculum.

Origin:	Number of fish per aquarium:	Diet:
Amazon region, Peru, Brazil	Minimum of 2 to 4	Dried food, small live and frozen foods

3 ROSY TETRA
Hyphessobrycon rosaceus
Size: **1¹/₂–2 in. (4–5 cm)**

The dorsal fin of the male is higher and more pointed than that of the female.

Origin:	Number of fish per aquarium:	Diet:
Guyana, Brazil	Minimum of 4 to 6	Dried food, small live and frozen foods

Head-and-tail-light tetra

This peaceful fish spends its time in or near plants, but it is capable of moving swiftly to obtain food. It prefers oxygenated water, with regular, partial changes to keep the nitrate level low. The female scatters her eggs on fine-leaved plants; the fry must be fed with brine shrimp nauplii.

A tetra for novices

Although difficult to breed, the lemon tetra (*Hyphessobrycon pulchripinnis*) is highly recommended for the beginner due to its robust nature and peaceful temperament. It does not exceed a length of 2 inches (5 cm) and prefers to live in a shoal. It readily accepts dried food.

4 HEAD-AND-TAIL-LIGHT TETRA
Hemigrammus ocellifer
Size: **2 in. (5 cm)**

This almost transparent characin is fairly easy to breed in fresh, acidic water.

Origin:	Number of fish per aquarium:	Diet:
North and west of South America	Minimum of 4	Dried food, small live and frozen foods

The Amazon aquarium

1 SERPAE TETRA
Hyphessobrycon eques

Family: Characidae
Size: 1¹/₂ in. (4 cm)

Origin:	Number of fish per aquarium:	Diet:
Paraguay basin	Shoal of 10 fish per 25-gal. (95 L) tank	Live foods and flakes

2 FLAME TETRA
Hyphessobrycon flammeus

Family: Characidae
Size: 1¹/₂ in. (4 cm)

Origin:	Number of fish per aquarium:	Diet:
Brazil	10 to 12 per 25-gal. (95 L) tank	Live foods and flakes; omnivorous

3 RUMMY-NOSE TETRA
Hemigrammus rhodostomus

Family: Characidae
Size: 2 in. (5 cm)

Origin:	Number of fish per aquarium:	Diet:
Amazon delta	Shoal of 10 to 15 fish per 25-gal. (95 L) tank	Live foods and flakes

4 SILVER HATCHETFISH
Gasteropelecus sternicla

Family: Gasteropelecidae
Size: 2–2¹/₂ in. (5–6.5 cm)

Origin:	Number of fish per aquarium:	Diet:
Southern Amazon basin	Shoal of 5 to 8 fish per 25-gal. (95 L) tank	Live food, flake or small floating granules

5 ZEBRA PLECO
Hypancistrus zebra

Family: Loricariidae
Size: 3–5 in. (7.5–12.5 cm)

Origin:	Number of fish per aquarium:	Diet:
Xingu River, Brazil	1 or 2 fish per 25-gal. (95 L) tank	Roots and live foods

6 GOLDIE PLECO
Scobinancistrus aureatus

Family: Loricariidae
Size: 8–12 in. (20–30 cm)

Origin:	Number of fish per aquarium:	Diet:
Xingu River, Brazil	1 or 2 fish per 50-gal. (190 L) tank, minimum	Roots and live foods

1

Black widow

This fish is undemanding with regard to water quality. It prefers a densely planted tank with moderate light. It likes to live in a small shoal and readily accepts the presence of other fish of the same size. It breeds in pairs or small groups, in acidic water of limited hardness. The female can lay several hundred eggs in a few hours; the fry hatch the next day and the parents must then be removed. The fry are initially dark but become paler; they readily feed on brine shrimp nauplii.

Black phantom tetra

This is one of the easiest characins to keep in an aquarium. It requires fairly soft, neutral or slightly acidic water. It likes swimming in moving water – near the filter outlet, for example. When it is frightened, this fish takes advantage of the vegetation to hide. Breeding is a tricky process that seems to depend on a good, varied diet and subdued lighting. The water must be filtered over peat to lower the pH to 6. Fry start swimming six days after hatching, but they often stay hidden and are slow to mature.

Diamond tetra

The diamond tetra likes swimming in a shoal in flowing water. It reaches maturity quickly on a diet based on small live prey, such as mosquito larvae, tubifex worms and brine shrimp.

Breeding is complicated but not impossible; soft, acidic water is required. The female scatters her eggs on fine-leaved plants. The eggs hatch two days later, and

2 BLACK PHANTOM TETRA
Hyphessobrycon megalopterus
Size: 1³/₄ in. (4.5 cm)
Its black coloring contrasts with that of brightly colored characins.

Origin:	Number of fish per aquarium:	Diet:
Eastern Brazil	2 to 4	Dried food, small live and frozen foods

1 BLACK WIDOW
Gymnocorymbus ternetzi
Size: 2 in. (5 cm)
There is a variety of this fish that has oversized fins and is therefore popular with breeders.

Origin:	Number of fish per aquarium:	Diet:
Amazon region, Brazil, Colombia	4 to 6	Dried food, small live and frozen foods

2

the parents must then be removed as they are likely to eat the eggs. The fry must be fed with brine shrimp nauplii. Other closely related species, *Moenkhausia sanctaefilomenae* (red-eye tetra) and *M. oligolepis* (glass tetra), have similar characteristics. Their fins are less prominent.

3 DIAMOND TETRA
Moenkhausia pittieri
Size: 2¹/₂ in. (6.5 cm)

This peaceful fish thrives in an aquarium shared with other characins. The male can be distinguished by its more pointed dorsal fin.

Origin:	Number of fish per aquarium:	Diet:
Venezuela	6	Dried food, small live and frozen foods

Getting around without vision

As its name indicates, the blind tetra (*Stygichthys typhlops*) has no eyes, as they are not needed in the deep waters where it lives. This does not prevent it from moving around and avoiding obstacles by using its lateral line system; and it detects its prey by an acute sense of smell. It is advisable to raise blind tetras in a shoal in a species tank, possibly in the company of a related species, the equally blind *Astyanax mexicanus* (right). Both of these fish prefer live food but they will accept dried food.

123

1 RUMMY-NOSE TETRA
Hemigrammus bleheri
Size: **2 in. (5 cm)**

This fish is distinguished by its bright red head, although it is often confused with a related species, *Petitella georgiae*.

Origin:	Number of fish per aquarium:	Diet:
Amazon region	4 to 6	Dried food, small live and frozen foods

2 GLASS BLOODFIN TETRA
Prionobrama filigera
Size: **2–2¹/₂ in. (5–6.5 cm)**

This little-known fish is almost transparent and creates a striking effect in a well-planted tank.

Origin:	Number of fish per aquarium:	Diet:
South America, particularly Brazil	Minimum of 6	Dried food, small live and frozen foods

Rummy-nose tetra

This is a placid, albeit shy fish that fits well in either a characin tank or a mixed community aquarium. It prefers water that is soft, acidic, brown (peat filtration) and frequently renewed. The female lays around 100 eggs, but the parents must be withdrawn immediately afterward. Not to be confused with *Hemigrammus rodostomus*, which has the same common name.

Glass bloodfin tetra

This fish tends to stay in the top part of the tank, where the water is constantly moving; it can even jump out of an aquarium if the tank is not completely covered. It tolerates neutral and moderately hard water and can live in the company of other species.

Bloodfin tetra

This fish can live in captivity for as long as 10 years. Its coloring is more vivid in water that is not too warm (72–73°F/22–23°C), but it can tolerate warmer conditions. Given sufficient space to swim undisturbed, it accepts the presence of other peaceful species of the same size in a aquarium.

The female often lays up to 500 eggs in the foliage of wispy plants. It is vital to remove the parents once the eggs have been laid, or they are liable to eat them.

Silver-tipped tetra

The silver-tipped tetra tends to take refuge anxiously in the plants, although it will swim in areas free of vegetation. In the wild, it lives in fast-flowing streams. Males have a white patch on the anal fin; this is yellow in females. Females lay up to 300 eggs, which the parents may sometimes eat.

3 BLOODFIN TETRA
Aphyocharax anisitsi
Size: **2 in. (5 cm)**

This fish is not fussy about water quality and is easy to keep in captivity.

Origin:	Number of fish per aquarium:	Diet:
Argentina	Minimum of 6	Dried food, small live and frozen foods

4 SILVER-TIPPED TETRA
Hasemania nana
Size: **2 in. (5 cm)**

This fish is ideally suited to a mixed community, in the company of peaceful fish.

Origin:	Number of fish per aquarium:	Diet:
Brazil	4 to 6	Dried food, small live and frozen foods

Penguin tetra

This is a placid, gregarious fish that needs to be kept in a shoal in a densely planted aquarium with subdued lighting. It can tolerate fairly hard water, but prefers it soft and slightly acidic, especially when breeding. Reproduction is fairly easy to achieve in a small tank with fine-leaved plants; introduce one male for every two or three females. One female can lay several hundred eggs (this often this happens in the morning) and these hatch the next day. The fry start off by eating infusorians, but after their third or fourth day of life they must be fed on brine shrimp nauplii. Also called hockey-stick tetra.

1 PENGUIN TETRA
Thayeria boehlkei

Size: **2 in. (5 cm)**

Penguin tetras adopt a distinctive posture, always swimming obliquely with their heads elevated.

Origin:	Number of fish per aquarium:	Diet:
Brazil, Peru	**4 to 6**	**Dried food, small live and frozen foods**

Marbled hatchetfish

This fish is very sociable and enjoys living in a shoal. As it lives just beneath the surface of the water, it can be raised with other tranquil species that occupy the bottom and midwater zones of a tank. Its ample pectoral fins allow it to jump several inches out of the water – so make sure that the aquarium is well covered. There are other, closely related (but unmarbled) species; these can be smaller, such as *Carnegiella marthae*, or bigger, such as those of the *Gasteropelecus* genus. They all behave in the same way. Little information is available about the breeding of these fish, but they appear to mate in groups.

2 MARBLED HATCHETFISH
Carnegiella strigata

Size: **2 in. (5 cm)**

This is one of several species of hatchetfish that live just below the surface and are liable to leap out of the tank.

Origin:	Number of fish per aquarium:	Diet:
Amazon region, Guyana	**4 to 6**	**Preference for small live and frozen foods, at the surface of the water**

Splash tetra

The males are generally more brightly colored than the females and their fins are bigger. This characin uses an unusual breeding method. The spawning pair jump out of the water again and again to place eggs under the leaf of a plant, and the male then splashes them regularly to keep them moist. When the eggs hatch, the fry fall into the water, where they must be fed brine shrimp nauplii. To foster this breeding behavior in captivity, provide a tank with a tight-fitting cover glass and suitable plants that overhang the water surface.

Golden pencilfish

This cousin of the splash tetra is easy to keep in captivity. It is somewhat timid and can be overwhelmed by bigger fish. Its coloring changes with the light: by day, stripes are visible on its flanks, but at night small vertical bars appear.

There are several closely related species available, including the one-lined and three-lined pencilfish.

3 SPLASH TETRA
Copella arnoldi

Size: male 3 in. (7.5 cm), female 2½ in. (6.5 cm)

Its pointed shape enables it to jump out of the water, and the tank must therefore be fully covered.

Origin:	Number of fish per aquarium:	Diet:
Guyana and the downstream area of the Amazon	1 pair	Dried food, small live and frozen foods

4 GOLDEN PENCILFISH
Nannostomus beckfordi

Size: 2 in. (5 cm)

The males are more pointed than the females and have small white patches on their flanks.

Origin:	Number of fish per aquarium:	Diet:
Parts of the Amazon region and Brazil	2	Preference for small live and frozen foods, but accepts dried food

Red piranha

The piranha needs a densely planted tank of at least 50 gallons (190 L) to feel at ease. It also prefers subdued lighting. Obviously, it can only live with other piranhas – any other fish will be considered prey (particularly if they are small). The water should be soft and acidic; regular, partial water changes are advisable. It rarely breeds in an aquarium. There are a few related species that sometimes appear on the market; they have the same requirements.

1 RED PIRANHA
Pygocentrus nattereri
Size: 8 in. (20 cm)

Although this fish has a bloodcurdling reputation in the wild, it proves quite placid in an aquarium. Nevertheless, it is advisable to keep your hands out of the water!

Origin:	Number of fish per aquarium:	Diet:
Amazon region	A few specimens	Live and frozen foods, animal flesh (meat, fish), small live fish

2 SILVER DOLLAR
Metynnis hypsauchen
Size: 4–6 in. (10–15 cm)

This relative of the piranha is less dangerous, as it is basically herbivorous.

Origin:	Number of fish per aquarium:	Diet:
South America	A few specimens	Plant-based foods

Silver dollar

If this fish is not fed with plant-based foods, it may attack the aquarium plants. Otherwise, it is peaceful, even fearful, and it is best to raise it with other placid species in a large tank. It is fairly awkward to breed (and requires extremely acidic water for this purpose). The female can lay over 1,000 eggs and the fry grow very quickly.

Red-eye characin

This fish sports bright colors under good conditions (neutral or slightly acidic water with negligible hardness). The male's anal fin is convex, with red, yellow and black stripes; the female's is pale with a black tip. Breeding requires soft water with a pH of 6–6.5. Females can lay up to 1,000 eggs, which hatch in about 36 hours. The fry start swimming two days later and grow quickly; they should be fed small live prey and fine powdered food.

Congo tetra

This fish needs a moderately planted tank, but it enjoys swimming and so should be given open space to swim freely. When mature, the male develops a long flowing dorsal fin; the central rays of the caudal fin are also extended. Companion fish kept with Congo tetras should be chosen carefully, because they could damage these elegant fins. The water should be soft and neutral, or slightly acidic. Females lay up to 300 eggs on fine-leaved plants after a vigorous spawning session that can be spurred by a partial water change or early morning sunshine striking the tank. The eggs hatch six days later, and the fry readily feed on brine shrimp nauplii.

3 RED-EYE CHARACIN
Arnoldichthys spilopterus
Size: **3 in. (7.5 cm)**

This is a lively but peaceful fish suited to moderately planted aquariums. It can be raised in a group with other placid species of the same size.

Origin:	Number of fish per aquarium:	Diet:
One of the rare African characins (West Africa, Lagos, Niger delta)	2	Omnivorous

4

4 CONGO TETRA
Phenacogrammus interruptus
Size: **male 3$\frac{1}{2}$ in. (9 cm), female 2 in. (5 cm)**

Mature males are easy to recognize, due to the outgrowth on the caudal fin. They sometimes fight among themselves.

Origin:	Number of fish per aquarium:	Diet:
Central African lakes and River Congo	6	Omnivorous

A Central American cichlid aquarium

1 REDHEAD CICHLID
Vieja synspilus

Size: 10–12 in. (25–30 cm)

Origin:	Number of fish per aquarium:	Diet:
Mexico, Belize, Guatemala	1 couple per 135-gal. (510 L) tank	Omnivorous

2 MIDAS CICHLID
Amphilophus citrinellus

Size: male 12 in. (30 cm), female 8 in. (20 cm)

Origin:	Number of fish per aquarium:	Diet:
Nicaragua	1 pair per 135-gal. (510 L) tank	Omnivorous

3 JAGUAR CICHLID
Nandopsis managuensis

Size: male 12 in. (30 cm), female 10 in. (20 cm)

Origin:	Number of fish per aquarium:	Diet:
Costa Rica, Nicaragua	1 pair per 135-gal. (510 L) tank	Omnivorous predator

4 PEARLSCALE CICHLID
Herichthys carpintis

Size: male 10 in. (25 cm), female 8 in. (20 cm)

Origin:	Number of fish per aquarium:	Diet:
Northern Mexico	1 pair per 80-gal. (300 L) tank	Vegetarian

5 YELLOW BELLY CICHLID
Cichlasoma salvini

Size: male 8 in. (220 cm), female 6 in. (15 cm)

Origin:	Number of fish per aquarium:	Diet:
Mexico, Belize, Guatemala	1 pair per 105-gal. (400 L) tank	Omnivorous

6 GUAYAS CICHLID
Cichlasoma festae

Size: male 16 in. (40 cm), female 12 in. (30 cm)

Origin:	Number of fish per aquarium:	Diet:
Ecuador	1 pair per 125-gal. (475 L) tank	Omnivorous

Cichlids

This family includes fish from South America, which live in soft, acidic water, and from Africa, particularly the great lakes to the east, where the water is hard and alkaline. Cichlids are famous for their distinctive, sometimes pugnacious behavior and their method of reproduction: some of them incubate the eggs in their mouths.

Angelfish

Several varieties of this fish have been bred for the hobby: golden, black, marbled and with sailfins. Whatever type you choose, make sure they only cohabit with placid species that will not attack their fins. Small fish must also be avoided, because angelfish are liable to eat them.

Angelfish like a tank with abundant plants, particularly vals (*Vallisneria* spp.), that can provide them with hiding places. They also require open water for swimming.

1

1

2

1 ANGELFISH
Pterophyllum scalare
Size: **6 in. (15 cm) long, almost as tall**
This is a visually striking fish much loved by aquarists!

Origin:	Number of fish per aquarium:	Diet:
Central Amazon region, Peru, Ecuador	4	Small live and frozen food, artificial food

Different varieties

Some varieties occur naturally (green discus, blue discus), while others (red, red spot, snakeskin, pigeon, turquoise, white) are the result of patient breeding in captivity, undertaken by professionals and enthusiasts all over the world. All these can be shown in competition – there is even a world championship.

Brown discus

This fish is characterized by its nine dark vertical stripes, of varying degrees of intensity. It is peaceful and swims slowly. It can be raised with smaller, compatible species from its native habitat, such as neons and *Corydoras* catfish.

In view of its size, it requires an aquarium that is fairly large, at least 80 gallons (300 L), and deep, 20–24 inches (50–60 cm). The decor should include bogwood and plants, as it spends much of its time in them.

Another discus

The Heckel discus (*Symphysodon discus*) is considerably less common than the brown discus, and more difficult to rear. It is distinguished by the presence of three dark, vertical lines.

Altum angelfish

Although it comes from the same region as the angelfish, the altum angelfish (*Pterophyllum altum*) is less common in aquariums. Its fins are more elongated, making it look bigger. The body has three sharply defined dark bands. It is difficult to breed.

2 BROWN DISCUS
Symphysodon aequifasciatus
Size: 8 in. (20 cm)

One of the most majestic of all freshwater aquarium fish, but also one of the most expensive!

Origin:	Number of fish per aquarium:	Diet:
Amazon, Orinoco	2 to 4	Preferably small live and frozen foods, but accepts dried food

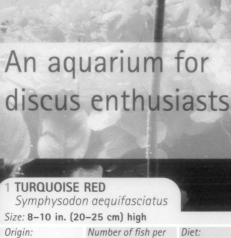

An aquarium for discus enthusiasts

1 TURQUOISE RED
Symphysodon aequifasciatus
Size: 8–10 in. (20–25 cm) high

Origin:	Number of fish per aquarium:	Diet:
Selected breeding	4 adults per 80-gal. (300 L) tank	Live and frozen food

2 PIGEON BLOOD
Symphysodon aequifasciatus
Size: 8–10 in. (20–25 cm) high

Origin:	Number of fish per aquarium:	Diet:
Selected breeding	4 adults per 80-gal. (300 L) tank	Live and frozen food

3 PIGEON SNAKE
Symphysodon aequifasciatus
Size: 8–10 in. (20–25 cm) high

Origin:	Number of fish per aquarium:	Diet:
Selected breeding	4 adults per 80-gal. (300 L) tank	Live and frozen food

4 COBALT
Symphysodon aequifasciatus
Size: 8–10 in. (20–25 cm) high

Origin:	Number of fish per aquarium:	Diet:
Selected breeding	4 adults per 80-gal. (300 L) tank	Live and frozen food

5 SNAKESKIN
Symphysodon aequifasciatus
Size: 8–10 in. (20–25 cm) high

Origin:	Number of fish per aquarium:	Diet:
Selected breeding	4 adults per 80-gal. (300 L) tank	Live and frozen food

6 RED MARLBORO
Symphysodon aequifasciatus
Size: 8–10 in. (20–25 cm) high

Origin:	Number of fish per aquarium:	Diet:
Selected breeding	4 adults per 80-gal. (300 L) tank	Live and frozen food

Firemouth cichlid

This normally placid fish only becomes aggressive during breeding, so it can be kept with other fish of the same size. The male is recognizable by its more pointed dorsal and anal fins.

Females lay up to several hundred eggs on a hard surface, and then go on to keep watch over both eggs and fry.

Convict cichlid

Because of its highly territorial behavior, especially when breeding, this cichlid must be housed on its own or with other cichlids from its native habitat in a fairly large tank. Plants are to be avoided – the decor should consist of various rocks and branches that form hideouts. Breeding is simple. The parents watch over the eggs, which are laid on a stone. They then guard the fry in a small hollow dug out of the substrate.

1 FIREMOUTH CICHLID
Thorichthys meeki
Size: **5 in. (12.5 cm)**

This fish displays the bright red skin on its throat to intimidate other fish.

Origin:	Number of fish per aquarium:	Diet:
Central America, Mexico	2 to 4	Prefers small live and frozen food, but accepts dried food

2 CONVICT CICHLID
Archocentrus nigrofasciatus (Cichlasoma nigrofasciatum)
Size: **6 in. (15 cm)**

This fish defends its territory ferociously, particularly during the breeding period.

Origin:	Number of fish per aquarium:	Diet:
Central America	2 to 4	Prefers small live and frozen food, but accepts dried food

Ram

This is a peaceful fish that appreciates a densely planted tank with plenty of nooks and crannies to hide in. The male is distinguished by the elongated stripes at the base of its dorsal fin. Females lay their eggs on a stone, secretly; the eggs are small but numerous (a few hundred).

Oscar

The oscar is fairly placid but must only be kept with fish of the same size. The juveniles have a pale, marbled brown coloring, but they already possess the eye-spot on the caudal peduncle that is typical of this species.

An oscar can become docile and even eat directly from its keeper's hand. It is not overly difficult to breed these cichlids. Females lay up to several hundred eggs, sometimes as many as a thousand, on a flat stone. The fry grow quickly. Several color varieties of this fish have been bred, including golden, bronze and red.

3 RAM
Mikrogeophagus ramirezi (Apistogramma ramirezi)
Size: 2¹/₂ in. (6.5 cm)
This small cichlid is peaceful unless kept with members of its own species. A golden variety is also available.

Origin:	Number of fish per aquarium:	Diet:
Amazon region	1 pair	Small live and frozen food, dried food

4 OSCAR
Astronotus ocellatus
Size: up to 14 in. (35 cm)
Being one of the biggest cichlids, this fish requires at least a 105-gal. (400 L) tank.

Origin:	Number of fish per aquarium:	Diet:
Amazon region	1 or 2	Live and frozen food, meat, dried food

A South American cichlid aquarium

1 PANDA DWARF CICHLID
Apistogramma nijsseni
Size: male 2–3 in. (5–7.5 cm), female 1½ in. (4 cm)

Origin:	Number of fish per aquarium:	Diet:
Peru	1 pair per 13-gal. (50 L) tank	Omnivorous; live food

2 RED-LINED DWARF CICHLID
Apistogramma hongsloi
Size: male 2½–3 in. (6.5-7.5 cm), female 1½-2 in. (4-5 cm)

Origin:	Number of fish per aquarium:	Diet:
Colombia	1 male and 3 females per 25 gal. (95 L)	Omnivorous

3 AGASSIZ'S DWARF CICHLID
Apistogramma agassizii
Size: male 2½ in. (6.5 cm), female 1½ in. (4 cm)

Origin:	Number of fish per aquarium:	Diet:
Brazil	1 male and 3 females per 25 gal. (95 L)	Omnivorous

4 PEARL CICHLID
Geophagus brasiliensis
Size: 6–10 in. (15–25 cm)

Origin:	Number of fish per aquarium:	Diet:
Brazil	1 pair per 80-gal. (300 L) tank	Omnivorous

5 RED-HUMP EARTHEATER
Geophagus steindachneri
Size: 6 in. (15 cm)

Origin:	Number of fish per aquarium:	Diet:
Colombia	1 pair per 50-gal. (190 L) tank housing male cichlids	Omnivorous

6 BOLIVIAN BUTTERFLY CICHLID
Mikrogeophagus altispinosus
Size: 6 in. (15 cm)

Origin:	Number of fish per aquarium:	Diet:
Brazil, Bolivia	1 pair per 25-gal. (95 L) tank	Omnivorous

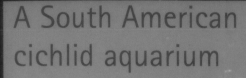

Lake Malawi cichlids

The fish featured on these two pages live only in Lake Malawi in Central Africa. They are often aggressive and require a tank containing rocks but no plants, as they will destroy them. They enjoy particularly hard and alkaline water conditions.

Golden mbuna
(Malawi golden cichlid)

This cichlid is aggressive and territorial both toward members of its own species and other fish. It requires a tank with rocky decor and plenty of hiding spots. The female incubates the eggs in her mouth and protects the fry in the same way.

Blue dolphin (Hump-head)

This sizable cichlid, relatively placid compared to others, needs a large tank of at least 105 gallons (400 L), where it can coexist with other species from the same lake. It will choose its territory among a pile of rocks. During breeding, the female incubates the eggs in her mouth.

2 BLUE DOLPHIN (HUMP-HEAD)
Cyrtocara moorii
Size: **10 in. (25 cm)**

As the name hump-head suggests, this cichlid is distinguished by the protuberance on its head.

Origin:	Number of fish per aquarium:	Diet:
Lake Malawi	1 pair	Small live and frozen food, artificial food

2

140

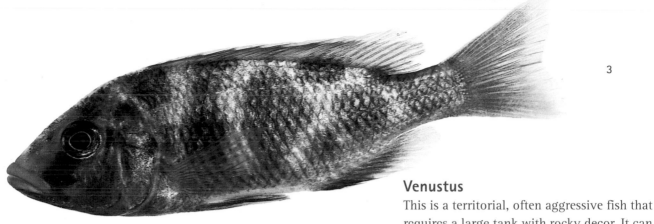

3

Venustus

This is a territorial, often aggressive fish that requires a large tank with rocky decor. It can coexist with other robust species of the same size. It is easy to breed in alkaline water. The female incubates eggs in her mouth for about three weeks. Once the fry are born, they return to their mother's mouth until they measure $1/2$ inch (1.25 cm). Males are more colorful than females.

3 VENUSTUS
Nimbochromis venustus (Haplochromis venustus)
Size: **8 in. (20 cm)**
The male is dark with white bands; the female is pale with yellow areas and two dark bands.

Origin:	Number of fish per aquarium:	Diet:
Lake Malawi	1 male per 3 or 4 females	Dried food, small live and frozen food

Zebra mbuna

This territorial fish can sometimes become aggressive. It needs at least a 50 gallon (190 L) tank, well stocked with hiding places and rocks that allow it to mark out its territory. The most common form of this cichlid is dark blue with vertical black bands, but there are other varieties, in pale blue and even orange.

1 GOLDEN MBUNA (MALAWI GOLDEN CICHLID)
Melanochromis auratus (Pseudotropheus auratus)
Size: **6 in. (15 cm)**
The male is dark with white bands; the female is pale with yellow areas and two dark bands.

Origin:	Number of fish per aquarium:	Diet:
Lake Malawi	1 pair, or 1 male and a few females	Food based mainly on plants, but also small live and frozen food, dried food

4 ZEBRA MBUNA
Metriaclima zebra (Pseudotropheus zebra)
Size: **6 in. (15 cm)**
The female incubates eggs in her mouth for three weeks. Keep her in a separate breeding tank during this period.

Origin:	Number of fish per aquarium:	Diet:
Lake Malawi	1 male per several females	Mainly plants and dried plant-based food, also small live and frozen food

A Lake Malawi cichlid aquarium

1 YELLOW LAB
Labidochromis caeruleus

Size: 10–12 in. (20–30 cm)

Origin:	Number of fish per aquarium:	Diet:
North of Lake Malawi	4 per 50-gal. (190 L) tank	Omnivorous

2 LEMON JAKE
Aulonocara jacobfreibergi

Size: 4–5 in. (10–12.5 cm)

Origin:	Number of fish per aquarium:	Diet:
Rocky shores of Lake Malawi	1 pair per 50-gal. (190 L) tank	Omnivorous

3 FLAVESCENT PEACOCK
Aulonocara stuartgranti

Size: 4–5 in. (10–12.5 cm)

Origin:	Number of fish per aquarium:	Diet:
Shores of Lake Malawi	1 pair per 50-gal. (190 L) tank	Omnivorous

4 COPADICHROMIS BORLEYI
Copadichromis borleyi

Size: 5–6 in. (12.5–15 cm)

Origin:	Number of fish per aquarium:	Diet:
Shores of Lake Malawi	1 pair per 80-gal. (300 L) tank	Omnivorous

5 CYNOTILAPIA AFRA
Cynotilapia afra

Size: 5–5$^{1}/_{2}$ in. (12.5–14 cm)

Origin:	Number of fish per aquarium:	Diet:
Shores of Lake Malawi	1 pair per 80-gal. (300 L) tank	Omnivorous

6 FOSSOROCHROMIS ROSTRATUS
Fossorochromis rostratus

Size: 10–12 in. (25–30 cm)

Origin:	Number of fish per aquarium:	Diet:
Lake Malawi	1 male and 2 females in minimum 135-gal. (510 L) tank	Omnivorous

Lake Tanganyika cichlids

Cichlid species from Lake Tanganyika are more placid than those from Lake Malawi; most of them can live together in an aquarium. They like hard, alkaline water and tend not to damage plants.

Frontosa

This large species requires at least a 105-gallon (400 L) tank furnished with rocky decor. It can coexist with other tranquil species. The female only lays a few dozen, fairly large eggs, which she incubates in her mouth for about a month.

1 FRONTOSA
Cyphotilapia frontosa
Size: **up to 14 in. (35 cm)**

The adult is distinguished by the bulge on its forehead.

Origin:	Number of fish per aquarium:	Diet:
Lake Tanganyika	1 pair	Small live and frozen food, dried food

2 GOLDEN JULIE
Julidochromis ornatus
Size: **3–4 in. (7.5–10 cm)**

This is a peaceful, territorial fish. Pairs remain faithful to each other.

Origin:	Number of fish per aquarium:	Diet:
Lake Tanganyika	1 pair	Small live and frozen food, dried food

Brichardi

The brichardi, or fairy cichlid, displays fascinating behavior patterns. It is basically peaceful but grows aggressive and territorial during the breeding period. It requires rocky decor with hideaways where it can lay its eggs. The male guards this area while the female takes care of both the eggs and the fry, which start swimming about a week after they hatch. The female can lay more eggs a month later.

Golden julie

Rocks forming small caves are essential for these fish, as they will use them when laying their eggs. There are several related species that differ in their coloring: *Julidochromis regani* (four brown horizontal lines), *J. marlieri* (brown-and-white checkerboard), *J. transcriptus* (a pattern of thick black bands) and *J. dickfeldi* (two dark bands, fins with blue borders).

Dwarf shell-dweller

This fish is unusual in that it lives in the empty shell of a gastropod native to the lake. So each fish must be provided with a fairly large shell (that of an edible snail or marine gastropod); it will use the shell to hide in when it is frightened, as well as to

reproduce. The parents take care of the fry, which move to a shell of their own when they are ¹/₂ inch (1.25 cm) long.

3 BRICHARDI CICHLID
Neolamprologus brichardi
Size: **4 in. (10 cm)**

This species lives in a group, often with fry from successive broods.

Origin:	Number of fish per aquarium:	Diet:
Lake Tanganyika	6 to 8	Small live and frozen food, dried food

4 DWARF SHELL-DWELLER
Neolamprologus brevis
Size: **1¹/2 in. (4 cm)**

A small, very peaceful cichlid that can live in a group and with other placid species.

Origin:	Number of fish per aquarium:	Diet:
Lake Tanganyika	4	Artificial food, small live and frozen food

145

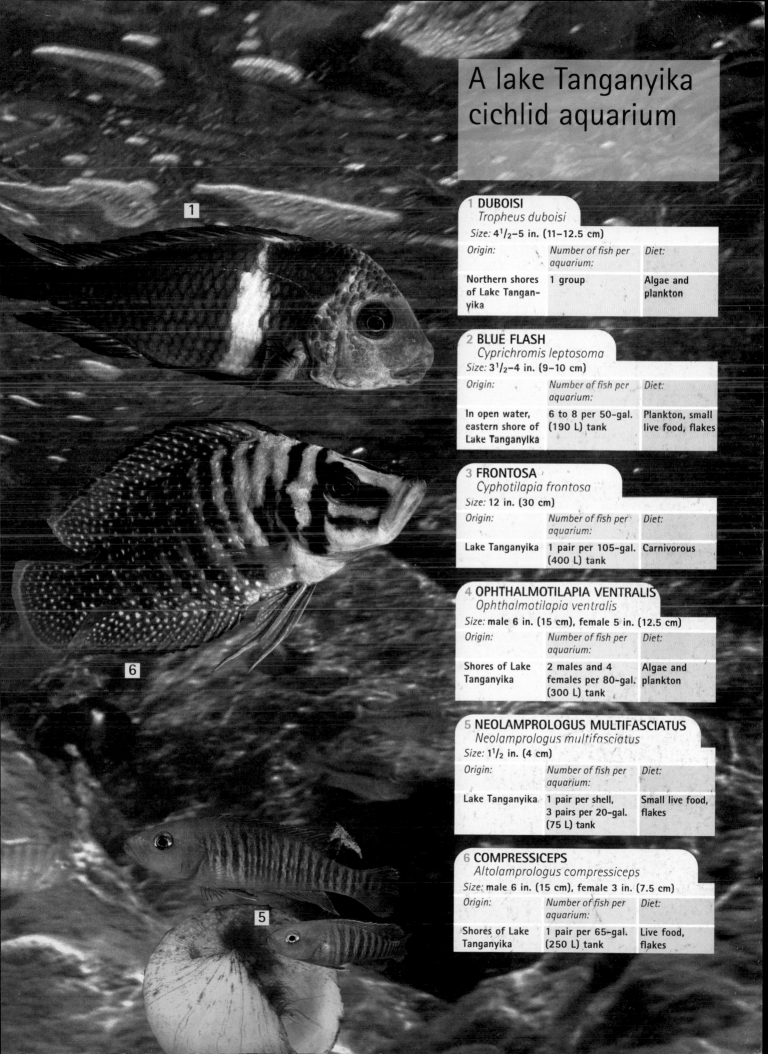

A lake Tanganyika cichlid aquarium

1 DUBOISI
Tropheus duboisi

Size: 4¹/₂–5 in. (11–12.5 cm)

Origin:	Number of fish per aquarium:	Diet:
Northern shores of Lake Tanganyika	1 group	Algae and plankton

2 BLUE FLASH
Cyprichromis leptosoma

Size: 3¹/₂–4 in. (9–10 cm)

Origin:	Number of fish per aquarium:	Diet:
In open water, eastern shore of Lake Tanganyika	6 to 8 per 50-gal. (190 L) tank	Plankton, small live food, flakes

3 FRONTOSA
Cyphotilapia frontosa

Size: 12 in. (30 cm)

Origin:	Number of fish per aquarium:	Diet:
Lake Tanganyika	1 pair per 105-gal. (400 L) tank	Carnivorous

4 OPHTHALMOTILAPIA VENTRALIS
Ophthalmotilapia ventralis

Size: male 6 in. (15 cm), female 5 in. (12.5 cm)

Origin:	Number of fish per aquarium:	Diet:
Shores of Lake Tanganyika	2 males and 4 females per 80-gal. (300 L) tank	Algae and plankton

5 NEOLAMPROLOGUS MULTIFASCIATUS
Neolamprologus multifasciatus

Size: 1¹/₂ in. (4 cm)

Origin:	Number of fish per aquarium:	Diet:
Lake Tanganyika	1 pair per shell, 3 pairs per 20-gal. (75 L) tank	Small live food, flakes

6 COMPRESSICEPS
Altolamprologus compressiceps

Size: male 6 in. (15 cm), female 3 in. (7.5 cm)

Origin:	Number of fish per aquarium:	Diet:
Shores of Lake Tanganyika	1 pair per 65-gal. (250 L) tank	Live food, flakes

Krib

This peaceful, sociable fish likes soft, slightly acidic water. The eggs are laid in a hideout that is guarded by the male, while the female looks after the eggs (there are usually more than 100). The incubation period lasts for three days, and the female only reappears a week after the hatching, accompanied by the fry. The small family is then protected by the male.

1 KRIB
Pelvicachromis pulcher (Pelmatochromis pulcher)
Size: **4 in. (10 cm)**

The female is distinguishable by her bright red, rounded stomach. The male has more pointed anal and dorsal fins.

Origin:	Number of fish per aquarium:	Diet:
West Africa, particularly Nigeria	1 pair	Dried food, small live and frozen food

Red jewel cichlid

The red jewel cichlid requires a tank with rocky decor and soft, acidic water. Its red color intensifies during the breeding period. When a pair bond is formed, the fish mark out their territory and actively defend it against all other fish. The eggs are laid on a stone and the fry emerge two days later.

2 RED JEWEL CICHLID
Hemichromis lifalili
Size: **5 in. (12.5 cm)**

The jewel cichlid previously available was *H. guttatus.* The one now on general sale is *H. lifalili.*

Origin:	Number of fish per aquarium:	Diet:
West Africa, particularly Nigeria	1 pair	Dried food, small live and frozen food

Lionhead cichlid

Despite its modest size, this fish can be aggressive when defending its territory. As it naturally lives in rivers, it prefers flowing water.

The female lays over 100 eggs in a cavity or on a concealed hard surface; the parents keep watch over them. They hatch a day later, and the parents' vigilance ceases once the fry start actively swimming.

Burton's mouthbrooder

This is a territorial fish that will only tolerate companions of the same size. Unlike most cichlids, it will not damage plants (but be sure to choose hardy varieties). Provide rocky decor with hiding places.

Breeding is fairly straightforward, with the female incubating the eggs in her mouth for two to three weeks, often hidden from view. Even after they have hatched, the fry may take refuge in the female's mouth at any sign of danger.

3 LIONHEAD CICHLID
Steatocranus casuarius
Size: **5 in. (12.5 cm)**

This cichlid is recognized by the lump on its forehead, which is more prominent in males.

Origin:	Number of fish per aquarium:	Diet:
Congolese rivers	1 pair	Small live and frozen food, dried food

4 BURTON'S MOUTHBROODER
Haplochromis burtoni (Astatotilapia burtoni)
Size: **6 in. (15 cm)**

The male is distinguished by yellow patches ringed with black on the anal fin.

Origin:	Number of fish per aquarium:	Diet:
Lakes Tanganyika and Kivu, East Africa	1 pair	Small live and frozen food, dried food

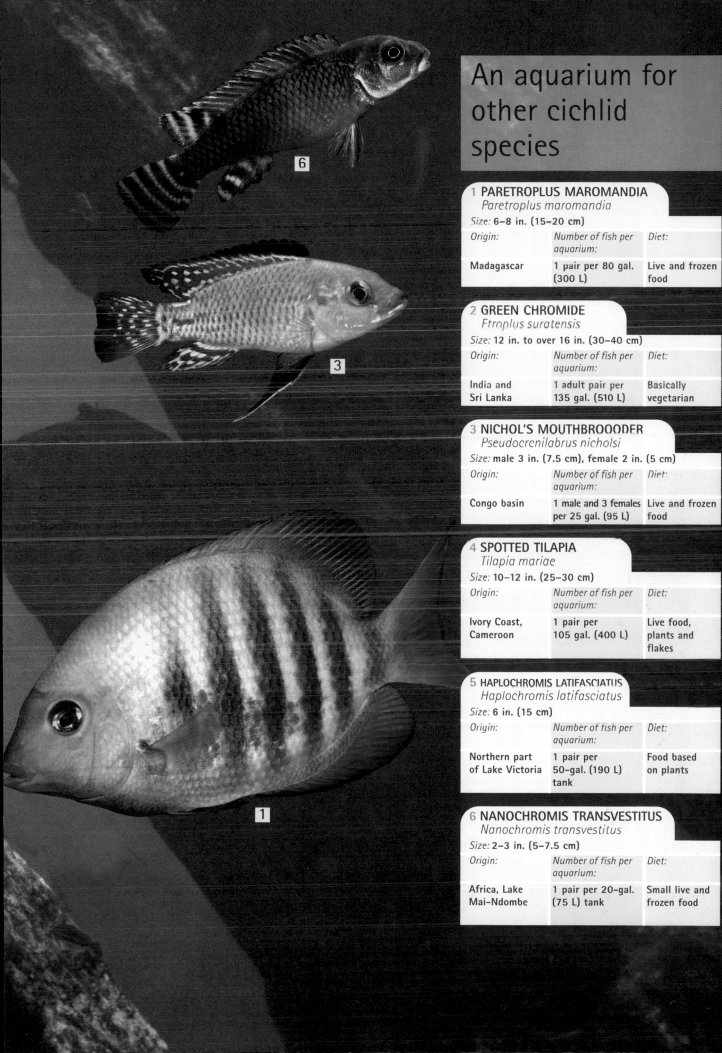

An aquarium for other cichlid species

1 PARETROPLUS MAROMANDIA
Paretroplus maromandia
Size: 6–8 in. (15–20 cm)

Origin:	Number of fish per aquarium:	Diet:
Madagascar	1 pair per 80 gal. (300 L)	Live and frozen food

2 GREEN CHROMIDE
Etroplus suratensis
Size: 12 in. to over 16 in. (30–40 cm)

Origin:	Number of fish per aquarium:	Diet:
India and Sri Lanka	1 adult pair per 135 gal. (510 L)	Basically vegetarian

3 NICHOL'S MOUTHBROOODER
Pseudocrenilabrus nicholsi
Size: male 3 in. (7.5 cm), female 2 in. (5 cm)

Origin:	Number of fish per aquarium:	Diet:
Congo basin	1 male and 3 females per 25 gal. (95 L)	Live and frozen food

4 SPOTTED TILAPIA
Tilapia mariae
Size: 10–12 in. (25–30 cm)

Origin:	Number of fish per aquarium:	Diet:
Ivory Coast, Cameroon	1 pair per 105 gal. (400 L)	Live food, plants and flakes

5 HAPLOCHROMIS LATIFASCIATUS
Haplochromis latifasciatus
Size: 6 in. (15 cm)

Origin:	Number of fish per aquarium:	Diet:
Northern part of Lake Victoria	1 pair per 50-gal. (190 L) tank	Food based on plants

6 NANOCHROMIS TRANSVESTITUS
Nanochromis transvestitus
Size: 2–3 in. (5–7.5 cm)

Origin:	Number of fish per aquarium:	Diet:
Africa, Lake Mai-Ndombe	1 pair per 20-gal. (75 L) tank	Small live and frozen food

Loaches

These are quite closely related to the cyrinids, but can be distinguished by their barbels, which they use to detect food, especially at night. As they are highly nocturnal, be sure to provide them with daytime shelters.

1

Clown loach

This highly gregarious loach tolerates members of its own species and other species from the *Botia* genus. Although largely nocturnal and somewhat timid, it can emerge by day to look for food. Otherwise, it remains hidden among plants and rocks. There is no record of it spawning in an aquarium.

A closely related species, *B. morleti*, does not exceed a length of 4 inches (10 cm) and has only one vertical black band on the caudal peduncle.

Yoyo loach

This loach needs a fine sandy substrate to prevent any damage to its barbels as it forages for food. It is an accommodating fish that tolerates the presence of other loaches. Although it is most active at night, it does not shrink from daylight. A related species, *Botia striata*, can be distinguished by the thin vertical stripes on its body. Neither of these species has been known to reproduce in an aquarium.

1 CLOWN LOACH
Botia macracanthus
Size: **6 in. (15 cm)**

This is the most popular of the loaches. It has not been known to breed in an aquarium.

Origin:	Number of fish per aquarium:	Diet:
Sumatra, Borneo	1 to 3, depending on the size of the tank	Dried food (including granules), small live and frozen food, snails

2 YOYO LOACH
Botia almorhae
Size: **3 in. (7.5 cm)**

When frightened or anxious, this fish makes a faint clicking noise with the spines on its operculum.

Origin:	Number of fish per aquarium:	Diet:
Northeast India, Bangladesh	1	Dried food (including granules), small live and frozen food

Dwarf botia

This fish is peaceful and active by day. It is averse to new water, so it must be put into an aquarium that has been in operation for a few months. Like other loaches, it has not been known to reproduce in captivity.

Coolie loach

Its snakelike appearance is very popular with aquarists. By day, the coolie loach hides in the decor; it comes out at night to feed.

The shape of its body allows it to weave its way throughout the aquarium and even slide out of it altogether – so make sure your lid fits tightly! It is also very adept at evading a landing net. Breeding is very difficult and unusual in an aquarium.

3 DWARF BOTIA
Botia sidthimunki
Size: 2 in. (5 cm)
The smallest of the loaches, it likes living in a group and is usually active during daylight

Origin:	Number of fish per aquarium:	Diet:
Northern India, Thailand	4 to 6	Dried food (including granules), small live and frozen food

4 COOLIE LOACH
Pangio kuhlii (Acanthophthalmus kuhlii)
Size: 4 in. (10 cm)
This name is used to describe several very closely related species that only differ in their color pattern.

Origin:	Number of fish per aquarium:	Diet:
Thailand, Malaysia, Sumatra, Java, Borneo	2 to 6	Dried food (including granules), small live and frozen food

4

Cyprinids

This family includes a large number of species, some of which are extremely popular, such as goldfish, danios, barbs and rasboras. Cyprinids are distinguished by their single dorsal fin and their scales. Some species have barbels that help them detect the wide range of foods that these omnivorous fish enjoy.

1

1 COMMON GOLDFISH
Carassius auratus
Size: **12 in. (30 cm)**
This is undoubtedly the best-known fish in the world.
There are several dozen varieties.

Origin:	Number of fish per aquarium:	Diet:
China, but introduced into all the temperate areas around the world	2	Dried food, small live and frozen foods

Common goldfish

The body of the common, or comet, gold-fish is thickset like that of a carp (a closely related species), but it does not share its barbels. Goldfish require a fairly spacious, well-filtered tank to live in captivity. (The claustrophobic confinement of the traditional glass bowl is quite unacceptable.) Heating serves no purpose, as goldfish tolerate a wide range of temperatures. They also like living in an outdoor pond.

Celestial goldfish

Its body is slimmer than that of the common goldfish and the tail fin is longer. It can be single-colored or two-toned. Celestials have "telescope" eyes that migrate upward as the fish mature. This means their vision is restricted, so keep them in a species tank where they need not compete for food with goldfish that have normal vision.

Breeding goldfish

This is easier to do in a pond than in an aquarium. A temperature of 66–68°F (19–20°C) is required. Goldfish reach maturity at two years of age. Males of reproductive age can often be recognized by the presence of tubercles on the head, which stimulate the females to lay eggs (sometimes several hundred at a time). The eggs are left on plants or an artificial medium, such as filter wool. They hatch in three or four days, and the fry start to feed two days later. In an aquarium, feed them on brine shrimp nauplii; in a pond, there is no need, as they will feed on animal plankton. The parents are liable to eat both eggs and fry if not removed after spawning.

Different varieties of goldfish

A classic goldfish has a metallic red or orange coloring, but other colors can be present, such as yellow, black, white or a mixture of these – in other words, no two goldfish are the same! The mutations that have long been specially bred are characterized by modifications to the form of the head, eyes, fins or the body. All these varieties can be crossbred, but in every case it is preferable to raise them in a pond.

Oranda

This short-bodied variety is characterized by the fleshy protuberances on top of the head, called a hood. There are many color forms, including the redcap oranda (left), which has a red hood and a silver body.

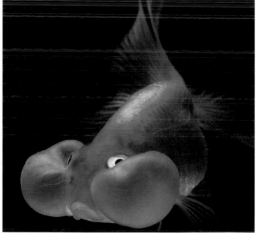

Shubunkin

The coloring is variable but is generally a mixture of white and orange-red with black patches. The dorsal and caudal fins can be elongated, but the caudal fin has only one lobe (unlike varieties such as the fantail).

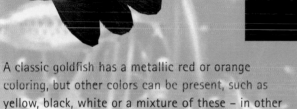

Bubble-eye

This extraordinary variety (above) has bag-like bubbles under the eyes. These are fluid-filled pouches of skin, and in the best examples should be of a similar size and shape. Bubble-eyes are poor swimmers and best kept in a separate aquarium away from other fancy goldfish.

An aquarium for coldwater fish

1 HI-FIN BANDED SHARK (BAT SUCKER)
Myxocyprinus asiaticus
Family: Cyprinidae
Size: **12–24 in. (30–60 cm)**

Origin:	Number of fish per aquarium:	Diet:
China	1 per 80 gal. (300 L)	Omnivorous

2 PEARLSCALE
Carassius auratus
Family: Cyprinidae
Size: **8–12 in. (20–30 cm)**

Origin:	Number of fish per aquarium:	Diet:
China	3 per 65 gal. (250 L)	Omnivorous

3 REDCAP ORANDA
Carassius auratus
Family: Cyprinidae
Size: **8–12 in. (20–30 cm)**

Origin:	Number of fish per aquarium:	Diet:
China	3 per 65 gal. (250 L)	Omnivorous

4 RYUKIN
Carassius auratus
Family: Cyprinidae
Size: **8–12 in. (20–30 cm)**

Origin:	Number of fish per aquarium:	Diet:
China	3 per 65 gal. (250 L)	Omnivorous

5 MOSQUITO FISH
Gambusia affinis
Family: Poeciliidae
Size: **male 1–1¹/₂ in. (2.5–4 cm), female 3 in. (7.5 cm)**

Origin:	Number of fish per aquarium:	Diet:
Mexico, southern United States	20 per 25-gal. (95 L) tank	Mosquito larvae, flakes

6 KOI CARP
Cyprinus carpio
Family: Cyprinidae
Size: **24–40 in. (60–100 cm)**

Origin:	Number of fish per aquarium:	Diet:
China	1 per 135 gal. (510 L)	Omnivorous

A barb for novices

The rosy barb *(Puntius conchonius, Barbus conchonius)* is recommended for beginners, as it is robust and undemanding with regard to water conditions. It grows to 3 or 4 inches (7.5 to 10 cm), and a variety with veiled fins is also bred. The male's coloring is more intense during the breeding period, which is similar to that of the cherry barb. To be sure of having a couple, it is advisable to acquire at least six specimens (this is true of most cyprinids).

Cherry barb

This fish owes its name to the color of the males, which is particularly vivid during the breeding period. The female has a dark band running from the mouth to the tail. She lays her eggs (sometimes upward of a hundred), in several stages, on the foliage of fine-leaved plants. Once spawning is over, the parents must be removed as they may eat the eggs. Eggs hatch in 24 hours; the fry feed on small live prey and fine dried food and grow quickly.

1 CHERRY BARB
Puntius titteya (Barbus titteya)
Size: **2 in. (5 cm)**
Peaceful, gregarious fish that likes living in a group in a densely planted tank.

Origin:	Number of fish per aquarium:	Diet:
Sri Lanka	Minimum of 4	Dried food, small live and frozen foods

1

Small and sociable

The golden barb (*Puntius gelius*) is one of the
smallest members of the Cyprinidae family –
it does not exceed 2 inches (5 cm) in length.
It prefers to live in a group of six to eight
specimens, in the company of other species of
a similar size. Its diet must obviously be compatible with the dimensions of its mouth:
small live prey and fine particles of dried food
are suitable. Females lay their eggs on a leaf
in soft, slightly acidic water; the parents must
then be removed, as they are liable to eat the
eggs, which hatch after 24 hours.

2 TIGER BARB
Puntius tetrazona (Barbus terazona)
Size: **3 in. (7.5 cm)**

This fish's boisterous behavior has given it a bad reputation. It is an
attractive option for an aquarium, however, if raised in a group.

Origin:	Number of fish per aquarium:	Diet:
Indonesia (Sumatra, Borneo)	Minimum of 6 to 8	Dried food, small live and frozen foods

Several varieties are bred, including
an albino and a green form. There is
also another closely related species, the
five-band barb (*Puntius pentazona*). Tiger
barbs are timid, so provide
a well-planted aquarium.
They are difficult to
breed and the fry
are hard to
raise.

Tiger barb

Avoid keeping tiger barbs with other fish
that have long fins, as this barb is
extremely likely to attack them.
Males display brighter colors
than females, particularly
during the breeding
period, when the pelvic
and dorsal fins turn
red. Soft water
with a neutral or
slightly acidic
pH is required
to induce
the female
to lay eggs.
The fry grow fairly
quickly; they reach a length
of $^1/_2$ inch (1.25 cm) in a
month, when their vertical
stripes appear.

2

hatch about 30 hours later. The fry grow quickly and reach full maturity in the space of a year.

Siamese flying fox

This is one of the most efficient devourers of the unwanted algae that grows in an aquarium. However, it must also be offered food with a vegetable base, otherwise it may damage aquarium plants. A closely related species, the flying fox (*Crossocheilus kalopterus*), is more brightly colored, but is considered less effective as an algae-eater. These two fish have not been known to reproduce in an aquarium.

Checker barb

This fish likes to have plenty of space to swim in a group – plants must be arranged on the sides or to the rear of the tank. It prefers slightly soft, acidic water.

Mature males are more colorful, with dark-edged orange fins. In the breeding period, they swim alongside the females, gently knocking their flanks to induce them to lay eggs. Offer the fish a varied diet.

1 CHECKER BARB
Puntius oligolepis (Barbus oligolepis)
Size: **3–3¹/₂ in. (7.5–9 cm)**
This barb can be recommended to novices, as it is easy to rear.

Origin:	Number of fish per aquarium:	Diet:
Indonesian rivers	Minimum of 6	Dried food, small live and frozen food

Gold barb

This is a very sociable fish that likes living in a group and is easy to breed in neutral water. The male is slimmer than the female, with a band on its flanks. The eggs are laid among fine-leaved plants and

2

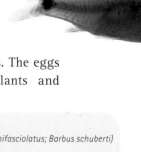

2 GOLD BARB (GREEN BARB)
Puntius semifasciolatus (Barbus semifasciolatus; Barbus schuberti)
Size: **3¹/₂ in. (9 cm)**
This fish is a mystery as its origins are unknown.
It is probably a hybrid developed by breeders.

Origin:	Number of fish per aquarium:	Diet:
Unknown	Minimum of 6	Dried food, small live and frozen foods

3 SIAMESE ALGAE-EATER
Crossocheilus siamensis
Size: **6 in. (15 cm)**

The tank must be tightly covered – this fish is a good jumper!

Origin:	Number of fish per aquarium:	Diet:
Thailand, Malaysia	1 (or more, depending on the size of the tank)	Plants, dried food predominantly based on plants

Red-tailed black shark

This is a fairly lively fish capable of defending its territory. It is basically gregarious, however, and tolerates the presence of other species. Spawning is rarely seen in an aquarium (in fact, professional breeders administer hormones to induce it to lay eggs). A related species, *Epalzeorhynchos frenatum*, is distinguished by its red fins; there is also an albino variety.

4 RED-TAILED BLACK SHARK
Epalzeorhynchos bicolor
Size: **6 in. (15 cm)**

Largely nocturnal, this fish hides by day.

Origin:	Number of fish per aquarium:	Diet:
Southeast Asia, particularly Thailand	1	Dried food based on plants, small live and frozen foods

want to add this popular fish to their collection immediately, but it is best to wait six to nine months so that the aquarium has a chance to settle down. The egg-laying process is distinctive: the female and male turn on their backs to lay and fertilize adhesive eggs on the lower surface of a large leaf. The characteristic triangular black patch appears about two weeks after birth.

Harlequin rasbora

This fish particularly appreciates small live prey. It prefers very soft, acidic water, especially during breeding. Most beginners

Red-tailed rasbora

The red-tailed rasbora needs an open area for swimming as it is continuously on the move. It is very sociable and can be raised along with danios or other rasboras. It likes a degree of darkness to spawn, together with fairly soft, neutral water.

Red-striped rasbora

This peaceful fish likes living in a group, with plant cover around the sides of the tank. Although slightly fearful, it accepts the company of other species. It breeds in very soft water that has been filtered through peat to make it acidic.

The males are thinner than the females. Eggs are laid on fine-leaved plants and hatch in 24 hours. The fry require very small pieces of food.

1 HARLEQUIN RASBORA
Rasbora heteromorpha
Size: 6 in. (15 cm)

This peaceful fish likes living in a group, in the company of other species from the same region.

Origin:	Number of fish per aquarium:	Diet:
Southeast Asia, particularly Thailand and Malaysia	6 to 8	Dried food, small live and frozen foods

2 RED-TAILED RASBORA
Rasbora borapetensis
Size: 2 in. (5 cm)

This little-known fish has a distinctive bright band on its flanks.

Origin:	Number of fish per aquarium:	Diet:
Thailand, Malaysia	6 to 8	Dried food, small live and frozen foods

Scissortail

If this fish is frightened, it tends to jump out of the aquarium, so provide a tight-fitting cover. Typically, it likes swimming in a space free of vegetation. It is prone to ich, especially if there is a drop in temperature or if it is stressed. Do not keep it with larger fish that may harass it. Very soft water and an increase in temperature stimulate spawning.

3 RED-STRIPED RASBORA
Rasbora pauciperforata
Size: 3 in. (7.5 cm)
This relatively uncommon fish is easily distinguished by its bright red stripe.

Origin:	Number of fish per aquarium:	Diet:
Sumatra, Malaysia	6 to 8	Dried food, small live and frozen foods

4 SCISSORTAIL
Rasbora trilineata
Size: 5 in. (12.5 cm)
One of the biggest rasboras, it owes its name to the way it moves its tail.

Origin:	Number of fish per aquarium:	Diet:
Malaysia, Sumatra, Borneo	4 to 6	Dried food, small live and frozen food

Zebra danio

This very active fish lives in a group and is constantly on the move around the aquarium – and therefore difficult to capture with a net! The zebra danio is an egglayer and easy to breed in a small tank with a deep coarse substrate, where the eggs will be protected from their parents' voracious appetite once they are laid.

The fry hatch in two days and will accept fine dried food. There is a veiltail form of this fish, with longer fins, which is very popular with breeders.

Giant danio

The common name for *Danio aequipinnatus (Devario aequipinnatus)* is an exaggeration, as it does not exceed 4 inches (10 cm) in length. It does, however, require more space than other danios, but like them it is peaceful and a good swimmer. The male has brighter colors (red on the belly). The eggs are laid in the morning in fine-leaved plants; the parents must then be removed. The fry grow quickly.

1

1 ZEBRA DANIO
Danio rerio (Brachydanio rerio)
Size: **2 in. (5 cm)**

This is one of the most popular of all aquarium fish, and is often the first one that novices acquire. It is also one of the easiest to keep.

Origin:	Number of fish per aquarium:	Diet:
Eastern coast of India	6 to 8	Dried food, small live and frozen foods

2 PEARL DANIO
Danio albolineatus (Brachyanio albolineatus)
Size: **2¹/₂ in. (6.5 cm)**

Its coloring is more diversified than that of the other danios.

Origin:	Number of fish per aquarium:	Diet:
Thailand, Malaysia, Sumatra	4 to 6	Dried food, small live and frozen foods

A mysterious danio

The leopard danio (*Danio rerio*) resembles a small trout. Its origins are unclear, but it is probably the result of a hybrid with the zebra danio, which has very similar characteristics. It is also available in the aquarium hobby as a veiltail form.

Pearl danio

The pearl danio is very active and swims the whole length of an aquarium. It is peaceful, lives in a group and is compatible with other danios.

A good diet and regular water changes enhance the intensity of its coloring. It is easy to breed under the influence of sunlight in a small tank equipped with fine-leaved foliage.

3 WHITE CLOUD MOUNTAIN MINNOW
Tanichthys albonubes

Size: 1¹/₂ in. (4 cm)

Easy to rear, this fish is recommended for beginners. It can tolerate a water temperature as low as 68°F (20°C).

Origin:	Number of fish per aquarium:	Diet:
China, near Canton	6 to 8	Dried food, small live and frozen foods

White Cloud Mountain minnow

This is a very sociable fish that lives in a group and can be raised in a small tank. Bear in mind that it cannot tolerate too much warmth for any length of time. Under good conditions, its coloring is very intense, earning it another common name: false neon. The males, more colorful and thinner than the females, put on a courting display to induce the latter to lay eggs on a fine-leaved plant. Some aquarists transfer their specimens to a garden pond in summer; here, they will reproduce easily.

An African aquarium

1 ORNATE CTENOPOMA
Microctenopoma ansorgii

Family: Anabantidae
Size: 3 in. (7.5 cm)

Origin:	Number of fish per aquarium:	Diet:
Congo Basin	1 pair per 25-gal. (95 L) tank	Freeze-dried food accepted, but preferably live foods

2 FEATHERFIN SYNODONTIS
Synodontis eupterus

Family: Mochokidae
Size: 6 in. (15 cm)

Origin:	Number of fish per aquarium:	Diet:
Chad and Niger basins	2 per 40-gal. (150 L) tank	Live foods; omnivorous

3 YELLOW-TAIL TETRA
Alestopetersius caudalis

Family: Alestidae
Size: 3–3$^{1}/_{2}$ in. (7.5–9 cm)

Origin:	Number of fish per aquarium:	Diet:
West Africa, Niger, Ghana, Togo	5 or 6 per 25-gal. (95 L) tank	Flakes and small live and frozen foods

4 SIX-BAR DISTICHODUS
Distichodus sexfasciatus

Family: Citharinidae
Size: 8–10 in. (20–25 cm)

Origin:	Number of fish per aquarium:	Diet:
Congo basin	2 or 3 for a tank of 160-gal. (600 L) or more	Plant-based foods, blanched lettuce and large flakes

5 PARATILAPIA POLLENI
Paratilapia polleni

Family: Cichlidae
Size: 8–12 in. (20–30 cm)

Origin:	Number of fish per aquarium:	Diet:
Lakes and marshes of Madagascar	1 pair per 105-gal. (400 L) tank	Large granules, insects and small fish

6 TANGANYIKA KILLIFISH
Lamprichthys tanganicanus

Family: Cyprinodontidae
Size: 5–6 in. (12.5–15 cm)

Origin:	Number of fish per aquarium:	Diet:
Lake Tanganyika	5 per 25-gal. (95 L) tank	Flakes, small live and frozen foods

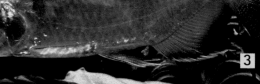

Killifish

This term includes species of fish that thrive in soft, acidic, stagnant water. Some of them live in temporary swamps that dry up at the end of the rainy season. Although the adult fish die as a result, the eggs they have laid survive in the ground and hatch when it starts raining again. In most species, the males are markedly more colorful than the females, with large dorsal and anal fins, and often lyre-shaped tail fins to impress potential mates. Most species can be housed and bred in small aquariums. They are fascinating subjects for more specialist fishkeepers.

Killi lyretail

This is one of the best-known species in this family. It lives for up to three years. Although it can live together with peaceful species of a similar size, it is best to keep it with other *Aphyosemion* species – although the females often look alike and there is a risk of hybridization. Breeding is fairly easy in a small tank if it is filtered with peat to obtain a pH of 6.5. The eggs hatch after two weeks and the fry reach maturity in a few months.

1 KILLI LYRETAIL
Aphyosemion australe
Size: 2 1/2 in. (6.5 cm)

The male is colorful and makes a striking aquarium fish. Cover the tank securely, as this species is capable of jumping out.

Origin:	Number of fish per aquarium:	Diet:
Cameroon, Gabon, Congo	1 male per 1 or 2 females	Dried food, small live and frozen foods

2 STRIPED PANCHAX
Aplocheilus lineatus
Size: 4 in. (10 cm)

This fish can behave aggressively toward members of its own species.

Origin:	Number of fish per aquarium:	Diet:
Central and southern India	1 male per 1 or 2 females	Preferably small live and frozen foods, will also accept dried food

Striped panchax

This fish mainly lives close to the surface and can jump out of the water if it becomes anxious. It has a voracious appetite and appreciates live foods. The anal and dorsal fins of the male are slightly larger than those in the female. Eggs are laid on plants and hatch in two weeks. The fry require very small particles of food, but grow quickly in good conditions. There are several closely related species, but these are rarely found in aquarium stores.

3

3 RED-TAIL NOTHO
Nothobranchius guentheri
Size: **2 in. (5 cm)**

There are several species of notho;
this is the most colorful.

Origin:	Number of fish per aquarium:	Diet:
East Africa	1 male per 1 or 2 females	Dried food, small live and frozen foods

Red-tail notho

This fish lives in small waterways or in swamps that evaporate in the dry season, while their eggs are preserved in the substrate. This remarkable process can be recreated in a small rearing tank with very soft water at pH 6.5. Peat should then be added so that the fish can bury their eggs in it. The eggs must be kept in the moist peat at a temperature of 61°F (16°C) for about two months. When water with identical characteristics is added, incubation begins and the fry emerge within two weeks.

4

4 BLACKFIN PEARLFISH
Austrolebias nigripinnis
Size: **2 in. (5 cm)**

The male is markedly more colorful than the female, with white patches against a bluish-brown background.

Origin:	Number of fish per aquarium:	Diet:
Argentina	1 male per 1 or 2 females	Dried food, small live and frozen foods

Blackfin pearlfish

Once this fish's eggs have been laid, they must be kept in moist peat for four months. There are a few other species related to *Austrolebias nigripinnis*, all from South America and all requiring the same treatment in regards to their eggs. The males are often aggressive with each other.

A killifish aquarium

1 BLUE GULARIS
Fundulopanchax sjostedti
Size: 5 in. (12.5 cm)

Origin:	Number of fish per aquarium:	Diet:
Nigeria, Cameroon	1 or 2 pairs in a 13-gal. (50 L) species aquarium	Live foods

2 TWO-STRIPED APHYOSEMION
Aphyosemion bitaeniatum
Size: 1 1/2–2 in. (4–5 cm)

Origin:	Number of fish per aquarium:	Diet:
Benin, Togo, Nigeria	1 or 2 pairs in a 13-gal. (50 L) species aquarium	Live foods

3 RED-STRIPED KILLIFISH
Aphyosemion striatum
Size: 2 in. (5 cm)

Origin:	Number of fish per aquarium:	Diet:
Gabon	1 or 2 pairs in a 13-gal. (50 L) species aquarium	Live foods

4 RED-CHINNED PANCHAX
Epiplatys dageti dageti
Size: 2–2 1/2 in. (5–6.5 cm)

Origin:	Number of fish per aquarium:	Diet:
Ivory Coast	1 or 2 pairs in a 13-gal. (50 L) species aquarium	Live foods

5 EGGERS' NOTHO
Nothobranchius eggersi
Size: 1 1/2–2 in. (4–5 cm)

Origin:	Number of fish per aquarium:	Diet:
Tanzania	1 or 2 pairs in a 13-gal. (50 L) species aquarium	Live foods

6 DOUANO KILLIFISH
Aphyosemion primigenium
Size: 1 1/2–2 in. (4–5 cm)

Origin:	Number of fish per aquarium:	Diet:
Gabon	1 or 2 pairs in a 13-gal. (50 L) species aquarium	Live foods

Rainbowfish (Melanotaeniidae family)

The fish from this family, popularly known as rainbowfish, originate from Australia and New Guinea. They like decidedly hard and alkaline water. They are omnivorous and accept all types of food.

Dwarf rainbowfish

This is the most common of the rainbowfish, but several very similar species are sold under this name. They will flourish as a small shoal in a community aquarium with a sandy substrate and feathery plants. Breeding is possible in hard, alkaline water. A female can lay up to 200 eggs on the fine leaves of plants; they stick to them with the help of a thin filament. The incubation lasts for a week and the fry grow slowly at first.

1

1 DWARF RAINBOWFISH
Melanotaenia maccullochi
Size: 3¹/₂ in. (9 cm)
This fish is a good swimmer that requires plenty of space.

Origin:	Number of fish per aquarium:	Diet:
New Guinea	4 to 6	Dried food, small live and frozen foods

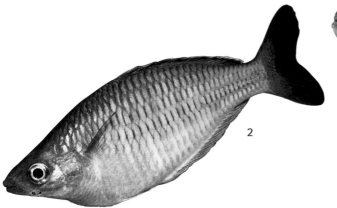

2

Boeseman's rainbowfish

This sociable fish lives in a small shoal and is an endless swimmer. The males are more brightly colored than the females, and their two dorsal fins are edged with white. This species can be bred successfully in an aquarium, but the fry grow slowly. Their bright coloring only starts to appear after about one year of life.

2 BOESEMAN'S RAINBOWFISH
Melanotaenia boesemani
Size: 4 in. (10 cm)
This is one of the most beautiful rainbowfish and is becoming increasingly available in aquarium stores.

Origin:	Number of fish per aquarium:	Diet:
New Guinea	4 to 6	Dried food, small live and frozen foods

Rainbowfish (Atherinidae family)

Very few species from this family live in fresh water. In fact, scientists disagree about their classification and sometimes assign them to other, little-known families. The ones shown here are also known as rainbowfish by aquarium hobbyists.

Celebes rainbowfish

This strong swimmer lives in open water and requires plenty of space. It prefers hard, alkaline water. It is a peaceful species that can be kept in a shoal. The male's second dorsal and anal fins are markedly more developed than those of the female. Breeding is not easy. It is necessary to add salt to the tank water to increase the hardness and raise the pH level to 7.5–8. The eggs are laid every day for a few days in fine-leaved plants and hatch eight to ten days later. The fry are often difficult to rear. They grow quite quickly and relish swimming in a current of water.

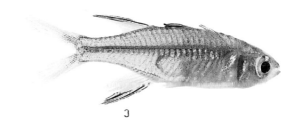

3

Madagascar rainbowfish

A good swimmer, it likes living in a small shoal in a fairly large tank. It is very gregarious and accepts the presence of species with the same temperament. Its coloring varies according to the conditions in which it is kept. Males are larger and more vivid than females. This species breeds in slightly alkaline water. The eggs fall to the bottom or among plants, and are in no danger of being eaten by the parents. Incubation lasts for one week, and the fry start swimming in search of food the day after they hatch. Fry will then take brine shrimp nauplii.

4

3 CELEBES RAINBOWFISH		
Macrosatherina ladigesi		
Size: 3 in. (7.5 cm)		
This brightly colored fish creates a striking display in the aquarium.		
Origin:	Number of fish per aquarium:	Diet:
Indonesia	4 to 6	Dried food, small live and frozen foods

4 MADAGASCAR RAINBOWFISH		
Bedotia geayi		
Size: 5 in. (12.5 cm)		
This peaceful and sociable fish can adapt to all types of water.		
Origin:	Number of fish per aquarium:	Diet:
Madagascar	4 to 6	Dried food, small live and frozen foods

An aquarium of curiosities

1 BUTTERFLYFISH
Pantodon buchholzi

Family: Pantodontidae
Size: 4 in. (10 cm)

Origin:	Number of fish per aquarium:	Diet:
Cameroon, Congo basin	3 or 4 per 50 gal. (190 L)	Insects, flakes

2 LEAF FISH
Monocirrhus polyacanthus

Family: Nandidae
Size: 4 in. (10 cm)

Origin:	Number of fish per aquarium:	Diet:
Peru	1 pair per 25 gal. (95 L)	Predator; live foods

3 SPOTTED GAR
Lepisosteus oculatus

Family: Lepisosteidae
Size: 24–40 in. (60–100 cm)

Origin:	Number of fish per aquarium:	Diet:
Lakes Erie and Michigan, Mississippi River	2 or 3 for a tank of 270+ gal. (1,000+ L)	Predator; live foods

4 LESSER SPINY EEL
Macrognathus aculeatus

Family: Mastacembelidae
Size: 12–14 in. (30–35 cm)

Origin:	Number of fish per aquarium:	Diet;
Southeast Asia	Single specimen per 320 gal. (1,200 L)	Carnivorous; live foods

5 SHOVEL-NOSE CATFISH
Sorubim lima

Family: Pimelodidae
Size: over 16 in. (40 cm)

Origin:	Number of fish per aquarium:	Diet:
Venezuela, Paraguay	2 per tank of 135+ gal. (510+ L)	Small live or frozen fish

6 WALKING CATFISH
Clarias batrachus

Family: Clariidae
Size: 12–16 in. (30–40 cm)

Origin:	Number of fish per aquarium:	Diet:
India, Thailand, Malaysia	2 or 3 per tank of 135+ gal. (510+ L)	Omnivorous; all foods

6

Catfish

This term is used to classify a number of fish belonging to a number of closely related families. Their main distinguishing feature are the barbels on their jaws, which help them to detect food, as well as play a tactile role. Provide these fish with a fine substrate to prevent any damage to these delicate organs.

Bronze cory

This fish is sociable and tolerates the presence of other corydoras in the same aquarium. It is generally indifferent to the water quality but prefers a soft, acidic water for breeding. The sexes are fairly easy to distinguish: the male has a more pointed dorsal fin and is shorter and thinner than the female, whose body cavity swells up with eggs. These are laid on a firm surface, be it a rock, plant or even a pane of the aquarium; they incubate in four or five days. The fry swim around less than a week later. Fry grow quickly and soon develop a taste for dried foods.

2

Peppered corys

Peppered cories can be bred in soft, acidic water. The patches on the male are more marked than those of the female. Females can lay several dozen eggs a day over the course of a week. The fry are fairly large at birth and accept brine shrimp nauplii. An albino variety is sometimes available in aquarium stores.

1 BRONZE CORY
Corydoras aeneus (Callichthyidae family)
Size: **3 in. (7.5 cm)**

This is the catfish most commonly found in aquariums. It is robust and ideal for beginners. The albino variety is shown below.

Origin:	Number of fish per aquarium:	Diet:
Northwestern South America, particularly Venezuela	2 to 6	Dried food (particularly granules), small live and frozen foods

2 PEPPERED CORY
Corydoras paleatus (Callichthyidae family)
Size: **3 in. (7.5 cm)**

Very gregarious fish that is fairly easy to breed.

Origin:	Number of fish per aquarium:	Diet:
Southeastern Brazil	2 to 6	Dried food, small live and frozen foods

1

Dwarf cory

This fish is active by day. Unlike other corydoras, it often swims in open water. It is peaceful and can be raised in a small tank in the company of other calm species. It is difficult to breed, as the female can only lay 25 to 30 eggs. The fry are fairly large at birth and grow quickly. The dwarf cory is sometimes confused with its equally small close relative, *C. pygmaeus*, although this species lacks the white marks around the black patch on the caudal peduncle.

3 DWARF CORY
Corydoras hastatus (Callichthyidae family)
Size: **1 in. (2.5 cm)**

This is the one of the smallest corydoras catfish.
It is rarely seen in aquarium stores.

Origin:	Number of fish per aquarium:	Diet:
Brazil (apparently in one river only)	4 to 6	Dried food, small live and frozen foods

4 THREE-LINED CORY
Corydoras trilineatus (Callichthyidae family)
Size: **2¹/₂ in. (6.5 cm)**

There are several varieties, depending on their geographical origin, but they differ only in the position of the black patches.

Origin:	Number of fish per aquarium:	Diet:
Ecuador, Peru, Colombia	2 to 6	Dried food (particularly granules), small live and frozen foods

Three-lined cory

Like all the corydoras, this species is peaceful and likes to live in a shoal. It is, however, considered more difficult to breed than the bronze cory and peppered cory. It looks very similar to *Corydoras julii* (known as the leopard cory); the coloring is practically identical. It is sometimes confused with other catfish species that have patches.

177

An oddballs aquarium

1 ELEPHANTNOSE
Gnathonemus petersii
Family: Mormyridae
Size: 6–8 in. (15–20 cm)

Origin:	Number of fish per aquarium:	Diet:
Nigeria, Cameroon	3 or 4 fish per 65 gal. (250 L)	Omnivorous; small live and frozen foods

2 RAPHAEL CATFISH
Platydoras costatus
Family: Doradidae
Size: 8–10 in. (20–25 cm)

Origin:	Number of fish per aquarium:	Diet:
Northern part of South America	1 or 2 per 65 gal. (250 L)	Omnivorous

3 SILVER AROWANA
Osteoglossum bicirrhosum
Family: Osteoglossidae
Size: 40–47 in. (100–118 cm)

Origin:	Number of fish per aquarium:	Diet:
Amazon Basin	1 per tank of 160+ gal. (600+ L)	All types of live foods

4 REEDFISH (ROPEFISH)
Erpetoichthys calabaricus
Family: Polypteridae
Size: 14–16 in. (35–40 cm)

Origin:	Number of fish per aquarium:	Diet:
Nigeria, Cameroon	3 or 4 per 50 gal. (190 L)	Various types of live foods

5 ATLANTIC MUDSKIPPER
Periophthalmus barbarus
Family: Gobiidae
Size: 6 in. (15 cm)

Origin:	Number of fish per aquarium:	Diet:
Coasts of West Africa	4 or 5 per 25-gal. (95 L) brackish tank	Live and frozen foods

6 GARFISH (NEEDLEFISH)
Xenentodon cancila
Family: Belontiidae
Size: 10–12 in. (25–30 cm)

Origin:	Number of fish per aquarium:	Diet:
Asia	2 or 3 per 50 gal. (190 L)	Predator; live and frozen foods

Porthole catfish

As this species is smaller than many other catfish, it is appropriate for a medium-sized tank; it will not take notice of other fish. It likes to swim in open spaces or move around on the substrate. As this catfish prefers eating at night, offer suitable foods when the lighting is turned off. It is not known to have bred in an aquarium.

Spotted pim

This species is basically peaceful but can sometimes attack small fish in the aquarium. It needs a large tank with plenty of space for swimming. There is no information about its breeding behavior, which appears to be unknown in an aquarium.

Whiptail catfish

This fish is fairly active and swims near the bottom during the day. Its mouth is set in a vertical position, allowing it to "stick" to the decor and grind the algae that it feeds on. The male can be identified by the bristles that appear on the side of the head during the breeding season. The female gener-

1 PORTHOLE CATFISH
Dianema longibarbis (Callichthyidae family)
Size: **4–6 in. (10–15 cm)**

This peaceful catfish is ideal for a community aquarium.

Origin:	Number of fish per aquarium:	Diet:
Peru	2	Dried food, small live and frozen foods

2 SPOTTED PIM
Pimelodus pictus (Pimelodidae family)
Size: **8 in. (20 cm) – often less in an aquarium**

The barbels of this nocturnal fish are often very prominent.

Origin:	Number of fish per aquarium:	Diet:
Amazon region	1	Dried food, small live and frozen foods

ally lays its eggs in a cave but will sometimes do so on the aquarium glass. The male cleans and guards the eggs. The fry need tiny foods such as infusorians when they first hatch.

Bristlenose catfish
Bristlenoses are territorial and may skirmish with other bottom-dwelling species. They like to graze on algae. If algae are not plentiful, then they may attack the aquarium plants, so provide plant-based foods to safeguard against this. The adult males can be distinguished from the females by the larger outgrowth of "bristles" – large skin outgrowths on and around the head. (A specimen without bristles is pictured on page 183.)

1

3 WHIPTAIL CATFISH
Sturisoma aureum (Loricariidae family)
Size: **8 in. (20 cm)**
This fish owes its name to its elongated caudal fin.

Origin:	Number of fish per aquarium:	Diet:
Colombia	1 or several, depending on the size of the tank	Plants, dried plant–based food, small live and frozen foods

4 BRISTLENOSE CATFISH
Ancistrus spp. (Loricariidae family)
Size: **5 in. (12.5 cm)**
There are several species of bristlenose, all fairly similar. Their behavior in the aquarium is identical.

Origin:	Number of fish per aquarium:	Diet:
Much of South America, particularly Brazil	2	Plants, dried plant–based food, small live and frozen foods

An aquarium of "useful" animals

1 JAPANESE SHRIMP
Caridina japonica
Size: 1–1¹/₂ in. (2.5–4 cm)

Origin:	Number of shrimp per aquarium:	Diet:
Asia	4 or 5 per 13 gal. (50 L)	Algae

2 MALAYAN SNAIL
Melanoides tuberculata
Size: ¹/₂–1¹/₂ in. (1.25–4 cm)

Origin:	Number of snails per aquarium:	Diet:
Tropical zones	5 per 3-gal. (11 L) tank	Algae, plant debris, leftover food

3 PLANORBIS
Planorbarius spp.
Size: ¹/₂–1 in. (1.25–2.5 cm)

Origin:	Number of fish per aquarium:	Diet:
East Asia	3 per 3-gal. (11 L) tank	Algae and plant debris

4 BEARDED CORY
Scleromystax barbatus
Family: Callichthyidae
Size: 4–5 in. (10–12.5 cm)

Origin:	Number of fish per aquarium:	Diet:
Brazil	2 per 25-gal. (95 L) tank	Omnivorous; live food

5 BRISTLENOSE CATFISH
Ancistrus temminckii
Family: Loricariidae
Size: 5–6 in. (12.5–15 cm)

Origin:	Number of fish per aquarium:	Diet:
Brazil	2 per 25-gal. (95 L) tank	Mainly plant-based food

6 SIAMESE ALGAE-EATER
Crossocheilus siamensis
Family: Cyprinidae
Size: 6 in. (15 cm)

Origin:	Number of fish per aquarium:	Diet:
Southeast Asia	2 per 25-gal. (95 L) tank	Plant-based food, algae

1 BUTTERFLY PLECO
Peckoltia pulcher (Dekeyseria pulcher) (Loricariidae family)
Size: **4 in. (10 cm)**

This territorial catfish will not cause any problems with other fish in a mixed community, but will quarrel with others of its kind.

Origin:	Number of fish per aquarium:	Diet:
Mainly Brazil	1	**Algae, dried and fresh plant-based foods**

2 CHOCOLATE-COLORED CATFISH
Rineloricaria lanceolata (Loricariidae family)
Size: **5–6 in. (12.5–15 cm)**

The long body shape and color patterns of this elegant catfish create almost perfect camouflage as it hugs the substrate.

Origin:	Number of fish per aquarium:	Diet:
Peru, Brazil, Bolivia, Paraguay	2	Plants, dried plant-based foods, small live and frozen foods

Butterfly pleco

Despite the small size of this fish, it needs space in the aquarium, otherwise it will certainly fight with other butterfly plecos. Ideally, it is best to keep just one individual in a community aquarium. It hides behind a rock or branch by day, then comes out in the evening to look for food, moving around with great energy. It will consume all the algae in the tank and still be hungry for more plant-based foods. There is little information on its breeding behavior.

Chocolate-colored catfish

This fish appreciates a current of water. It must be supplied with plant-based food to prevent it from attacking the vegetation in the tank. Mature males can be recognized by the large number of barbels on their head, the females by their darker color. The eggs are laid in caves and guarded by the parents; however, the fry are difficult to rear.

3 OTOCINCLUS
Otocinclus affinis (Loricariidae family)

Size: 1¹/₂ in. (4 cm)

This fish is an essential part of a mixed community tank, where it will keep algae in check and cause no problems with other fish.

Origin:	Number of fish per aquarium:	Diet:
Brazil	2	Plants, dried plant-based foods

Otocinclus

This fish is sensitive to the quality of the aquarium water, so efficient filtration and regular partial changes are essential to keep it happy. In the right conditions, these busy little catfish will constantly browse on the

4 SUCKERMOUTH CATFISH
Hypostomus plecostomus (Loricariidae family)

Size: 12 in. (30 cm)

This fish is hardy, peaceful and an excellent algae eater when small. It can grow large and disrupt the whole aquarium when mature.

Origin:	Number of fish per aquarium:	Diet:
Northern part of South America	1	Plants, dried plant-based foods

nuisance algae that grows on plant leaves and decor. The female lays her eggs on plant leaves and can take up to 72 hours to hatch. Feed the fry on very fine live foods as well as plant-based foods.

Suckermouth catfish

This fish is essentially nocturnal, although  it can be spotted by day eating tiny algae on the decor and plants, or even on the tank glass. It very rarely breeds in an aquarium. Think carefully before buying a suckermouth catfish unless you have a large tank; it may seem appealing when it is small, but it could become overwhelming when it grows to its full size of 12 inches (30 cm).

1

1 CHINESE ALGAE-EATER
Gyrinocheilus aymonieri (Gyrinocheilidae family)
Size: **12 in. (30 cm), less in an aquarium**

Despite its common name, it is not a very efficient algae-eating fish.
The largest specimens are extremely boisterous and often aggressive.

Origin:	Number of fish per aquarium:	Diet:
Thailand	1	Plants, dried plant-based food, small live and frozen foods

2 IRIDESCENT SHARK
Pangasius hypophthalmus (P. sutchi) (Pangasiidae family)
Size: **8–12 in. (20–30 cm) – often less in an aquarium**

Despite its name, this fish is placid and does not attack other fish.
It should be kept in a large aquarium with strong water currents.

Origin:	Number of fish per aquarium:	Diet:
Southeast Asia, particularly Thailand	1 or several small juveniles	Dried plant-based food, small live and frozen foods

Chinese algae-eater

Although the common name of this fish suggests it will be useful, it is far more likely to attach its sucker-shaped mouth to flat-bodied tankmates than control algae. When it grows big it can displace and knock over pieces of the decor as it is an energetic swimmer. There are no reports of it breeding in an aquarium.

Iridescent shark

This fish spends much of its time swimming, so it needs an open space free of plants and spacious aquarium for it to feel comfortable. It is omnivorous and accepts all types of food. This species does not appear to have bred in an aquarium.

2

Upside-down catfish

Its peculiar swimming style is suited to its habit of eating from the water's surface (mainly at night). Sociable and peaceful, it accepts all types of food but rarely breeds in an aquarium. Its eggs are laid in a cave. The fry swim normally at first and only adopt adult behavior patterns after two months. There are several closely related species in Africa, but they swim in the normal position. Of these, *Synodontis petricola*, with brown patches on a pale skin, is sometimes commercially available; it is slightly bigger and requires more space.

Glass catfish

This is one of the few aquarium fish with a transparent body. It lives in a group and does not enjoy the presence of overly frisky tankmates that could attack its long delicate barbels. It appreciates high-quality water, with a gentle current where it can swim horizontally. It rests in a sloping position, however, with its head higher than the tail. Dense planting will provide this fish with welcome cover. Keep them in a small shoal of at least four fish; single specimens will feel insecure and hide away, eventually dying through lack of food. Little is known about its breeding behavior.

3 UPSIDE-DOWN CATFISH
Synodontis nigriventris (Mochokidae family)

Size: male 3 in. (7.5 cm), female 4 in. (10 cm)

Do not worry if it swims with its belly facing upward – this is its normal posture; the fry swim right side up.

Origin:	Number of fish per aquarium:	Diet:
West Africa	1 or 2	Dried food, small live and frozen foods

4 GLASS CATFISH
Kryptopterus bicirrhis (Siluridae family)
Size: 5 in. (12.5 cm)

This distinctive fish is easy to recognize by its totally transparent body, with the spine and supporting bones clearly visible.

Origin:	Number of fish per aquarium:	Diet:
Eastern India, Thailand, Malaysia, Indonesia	4 to 6	Dried food, small live and frozen foods

A freshwater invertebrate aquarium

1 RAINBOW LAND CRAB
Cardisoma armatum
Size: 3 in. (7.5 cm) excluding legs

Origin:	Number of crabs per aqua-terrarium:	Diet:
West Africa, Malawi	1 per 3 ft. (1 m) of tank	Omnivorous

2 FIDDLER CRAB
Uca spp.
Size: 2–3 in. (5–7.5 cm)

Origin:	Number of crabs per aqua-terrarium:	Diet:
Asia	1 pair per 3 ft. (1 m) of tank	Omnivorous

3 GOLDEN APPLE SNAIL
Pomacea bridgesi
Size: 2–2$^1/_2$ in. (5–6.5 cm)

Origin:	Number of snails per aquarium:	Diet:
Tropical regions	4 per 25-gal. (95 L) tank	Algae, plants

4 RED CHERRY SHRIMP
Neocaridina denticulata
Size: 1–1$^1/_2$ in. (2.5–4 cm)

Origin:	Number of shrimp per aquarium:	Diet:
Asia	10 per 15-gal. (55 L) tank	Algae; scavenger

5 THAI SHRIMP
Macrobrachium spp.
Size: 2 in. (5 cm)

Origin:	Number of shrimp per aquarium:	Diet:
Southeast Asia	5 to 6 per 15 gal. (55 L)	Algae; scavenger

6 AUSTRALIAN BLUE CRAYFISH
Cherax tenuimanus
Size: 6 in. (15 cm)

Origin:	Number of crayfish per aquarium:	Diet:
Australia	1 per 25-gal. (55 L) tank	Scavenger

Livebearers

Livebearers are certainly the most popular of all aquarium fish. In the Central American Poeciliidae family, the males are easy to tell apart from the females, as the supporting rays on the anal fin are fused to form a reproductive organ, the gonopodium. During reproduction, the male uses this gonopodium to deposit his sperm in the female's genital orifice, in a kind of pseudo-copulation. The eggs grow and hatch inside the mother, and the fry are expelled alive and perfectly formed. They immediately start looking for food. Livebearers prefer hard, alkaline water.

1

1 GUPPY
Poecilia reticulata
Size: **2 in. (5 cm)**

This is undoubtedly the best-known of all aquarium fish. In the wild, its colors are considerably duller than in the numerous varieties bred by specialists.

Origin:	Number of fish per aquarium:	Diet:
Central America, Brazil	**At least 1 pair**	**Dried plant-based food, small live and frozen foods**

Guppy

The male is noticeably more colorful than the female, and its dorsal and caudal fins are larger. A female can produce about 20 to 40 offspring every four to six weeks. The fry grow rapidly and are mature within a few months. Guppies are gregarious fish, but it is advisable to keep them with placid species that will not be tempted to nip their flowing fins.

1

Guppies in competition

There are numerous varieties of guppies, which can be distinguished by their color (the cobra, for example), but the form of the caudal (tail) fin is the best indicator (fan-, delta- or lyre-shaped). Some enthusiasts patiently crossbreed their fish and display their finest specimens in competitions all over the world.

Black molly

The black molly is susceptible to certain diseases if it is not kept in hard or even slightly brackish water. This will be particularly apparent during the breeding period. Breeding is easy, although you should keep watch over the males when they fight for a female's attentions. Females give birth to several dozen fry.

Sailfin molly

This striking fish is very popular but can be difficult to keep properly. It needs the correct diet, sufficient space and a gentle flow of water. It appreciates warmth and hard water (you can even add a little aquarium salt) and is susceptible to disease

in poor-quality water. Males sometimes fight among themselves, but this fish is generally peaceful. A female can give birth to up to 100 large fry every two months. These can be fed from the start with fine dried or small live foods. They grow quickly and can start reproducing at nine months of age.

2 BLACK MOLLY
Poecilia sphenops
Size: male 3 in. (7.5 cm), female 3¹/₂–4 in. (9–10 cm)
This fish is completely black. There is a variety that sports a lyre-shaped tail fin.

Origin:	Number of fish per aquarium:	Diet:
Mexico, northern South America	At least 1 pair	Dried plant-based food, small live and frozen foods

3 SAILFIN MOLLY
Poecilia latipinna
Size: male 3 in. (7.5 cm), female 3¹/₂–4 in. (9–10 cm)
This molly differs from *P. sphenops* on account of its longer and higher dorsal fin. It can be gray speckled with black patches.

Origin:	Number of fish per aquarium:	Diet:
Southeastern United States, Mexico	At least 1 pair	Dried plant-based food, small live and frozen foods

1 YUCATAN SAILFIN MOLLY
Poecilia velifera

Size: 5 in. (12.5 cm)

The most imposing of the livebearers, there is a totally black form and a golden one (actually albino, with a lyre-shaped tail fin).

Origin:	Number of fish per aquarium:	Diet:
Mexico	At least 1 pair	Dried plant-based food, small live and frozen foods

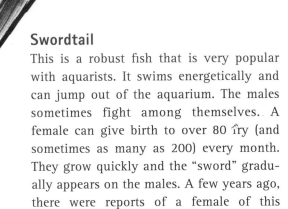

2

Yucatan sailfin molly

The males, easily distinguished by their prominent dorsal fin, sometimes fight among themselves to win a female. This molly needs water that is alkaline (pH 7.5 to 8) and hard (over 20°GH), so you can add 1 tablespoon (15 ml) of aquarium salt for every 5 gallons (19 L) of water to create ideal breeding conditions. Reproduction is straightforward, and one female can produce about 100 eggs every month. After being impregnated by a male, she can store his sperm to give birth several times in succession. The fry grow quickly and accept fine dried food as soon as they are born.

Swordtail

This is a robust fish that is very popular with aquarists. It swims energetically and can jump out of the aquarium. The males sometimes fight among themselves. A female can give birth to over 80 fry (and sometimes as many as 200) every month. They grow quickly and the "sword" gradually appears on the males. A few years ago, there were reports of a female of this species turning into a male. This phenomenon is not rare in fish – the scientific term is protogyny – but is not the case with the

1

2 SWORDTAIL
Xiphophorus helleri

Size: **male 4 in. (10 cm), female 5 in. (12.5 cm)**

Its name is derived from the extension at the base of the male's caudal fin. There are several varieties that differ in color and the length of the fins.

Origin:	Number of fish per aquarium:	Diet:
Central America	1 couple	Artificial plant-based food, small live and frozen prey

swordtail. Rather, the maturing process is delayed in some males, making them appear like females. Their gonopodia and swords start to grow after a while and they will then be able to reproduce.

Platy

The platy treats not only its own species but also others (particularly livebearers) with equanimity. It is liable to nibble plants if its diet is lacking in green foods. A female gives birth to several dozen fry, and these take four months to mature.

Variatus platy

The variatus platy (also called variable or variegated platy) has a placid temperament (the males do not fight each other). It is more prolific than the platy, as the female can produce a brood of 100 fry every month.

A livebearer for beginners

The platy is robust, prolific (up to a hundred fry) and easy to breed; it is therefore an excellent choice for the novice fishkeeper. Adult platies, being less than half the size of adult swordtails, make good community fish for those aquarists who only have space for a small aquarium.

In the wild the platy is pale yellow, but several varieties are bred: red (one of the most common), lemon (creamy yellow body) tuxedo (red with black patches), red wagtail (red to orange-yellow with black fins) and black (with a blue or green metallic sheen). The hi-fin platy has an overdeveloped dorsal fin; it can be found in any of the above colors. All these varieties can be crossbred, and the hybrids often combine features of both parents.

3

4

3 PLATY
Xiphophorus maculatus
Size: male 1–1^1/$_2$ in. (2.5–4 cm), female 2–2^1/$_2$ in. (5–6.5 cm)
The platy is one of the easiest livebearers to keep in captivity. There are several varieties.

Origin:	Number of fish per aquarium:	Diet:
Mexico, Guatemala, Honduras	At least 1 pair	Dried plant-based food, small live and frozen foods

4 VARIATUS PLATY
Xiphophorus variatus
Size: male 1–1^1/$_2$ in. (2.5–4 cm), female 2–2^1/$_2$ in. (5–6.5 cm)
This is a cousin of the platy and can be crossbred with it. The resulting offspring will be fertile.

Origin:	Number of fish per aquarium:	Diet:
Mexico, Guatemala	At least 1 pair	Dried plant-based food, small live and frozen foods

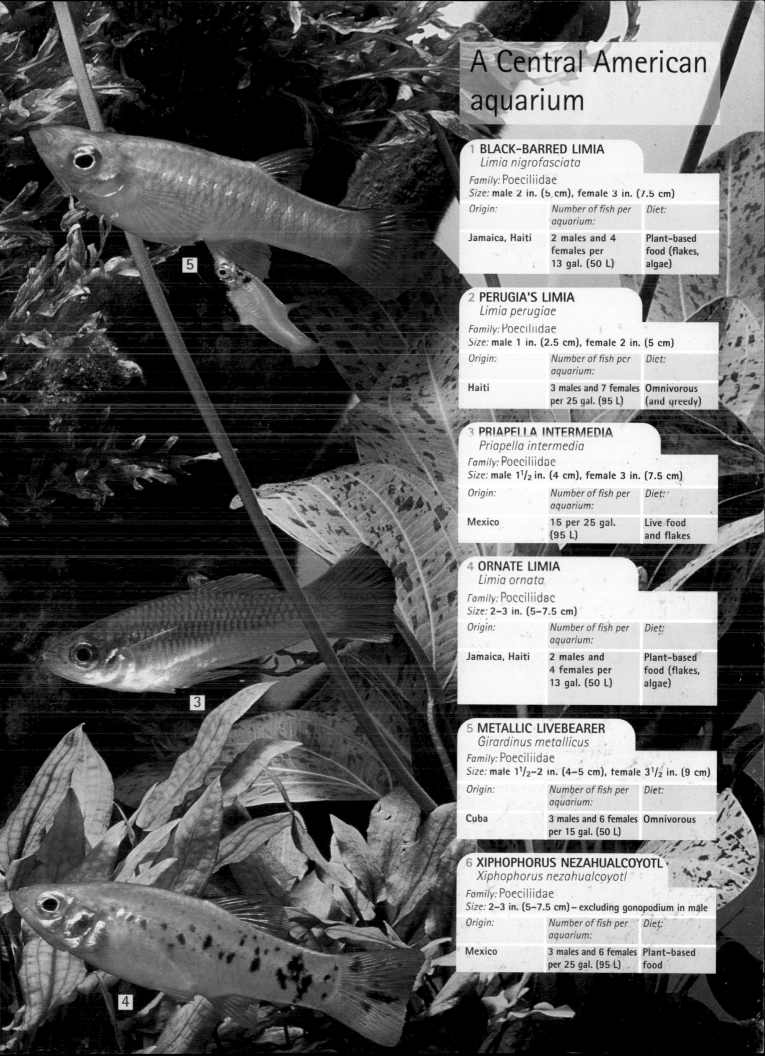

A Central American aquarium

1 BLACK-BARRED LIMIA
Limia nigrofasciata

Family: Poeciliidae
Size: male 2 in. (5 cm), female 3 in. (7.5 cm)

Origin:	Number of fish per aquarium:	Diet:
Jamaica, Haiti	2 males and 4 females per 13 gal. (50 L)	Plant-based food (flakes, algae)

2 PERUGIA'S LIMIA
Limia perugiae

Family: Poeciliidae
Size: male 1 in. (2.5 cm), female 2 in. (5 cm)

Origin:	Number of fish per aquarium:	Diet:
Haiti	3 males and 7 females per 25 gal. (95 L)	Omnivorous (and greedy)

3 PRIAPELLA INTERMEDIA
Priapella intermedia

Family: Poeciliidae
Size: male 1$^1/_2$ in. (4 cm), female 3 in. (7.5 cm)

Origin:	Number of fish per aquarium:	Diet:
Mexico	15 per 25 gal. (95 L)	Live food and flakes

4 ORNATE LIMIA
Limia ornata

Family: Poeciliidae
Size: 2–3 in. (5–7.5 cm)

Origin:	Number of fish per aquarium:	Diet:
Jamaica, Haiti	2 males and 4 females per 13 gal. (50 L)	Plant-based food (flakes, algae)

5 METALLIC LIVEBEARER
Girardinus metallicus

Family: Poeciliidae
Size: male 1$^1/_2$–2 in. (4–5 cm), female 3$^1/_2$ in. (9 cm)

Origin:	Number of fish per aquarium:	Diet:
Cuba	3 males and 6 females per 15 gal. (50 L)	Omnivorous

6 XIPHOPHORUS NEZAHUALCOYOTL
Xiphophorus nezahualcoyotl

Family: Poeciliidae
Size: 2–3 in. (5–7.5 cm) – excluding gonopodium in male

Origin:	Number of fish per aquarium:	Diet:
Mexico	3 males and 6 females per 25 gal. (95 L)	Plant-based food

Least killifish

The least killifish prefers hard, alkaline water. Its coloring is enhanced by a diet based on live food. The female gives birth to relatively few fry, and these need plants to hide within. The fry feed on brine shrimp nauplii and fine dried food.

1 LEAST KILLIFISH
Heterandria formosa
Size: **male 1 in. (2.5 cm), female 2 in. (5 cm)**
This diminutive livebearer is little known. It is one of the smallest fish in the world.

Origin:	Number of fish per aquarium:	Diet:
Southeastern United States	1 pair	Dried food, small live and frozen food

Red-tailed goodeid

The males are somewhat aggressive, both among themselves and toward other species; they can be distinguished by the orange color at the rear of the body and the slightly modified anal fin (although this is more pronounced in guppies, platies and swordtails). Breeding requires water that is hard (add a little salt) and alkaline (pH between 7 and 8). The females give birth about every two months to approximately 50 fry, which are very slow growing.

2 RED-TAILED GOODEID
Xenotoca eiseni
Size: **2–3 in. (5–7.5 cm)**
This member of the Goodeidae family is rarely seen in an aquarium.

Origin:	Number of fish per aquarium:	Diet:
Mexico	At least 1 pair	Dried food, small live and frozen foods

Butterfly goodeid

The males are more colorful and the anal fin is only slightly modified. The females cannot store sperm, so must mate each time before they produce a brood of young.

Halfbeak

This fish can be recognized by the distinctive form of its lower jaw, which is far more developed than the upper one. This allows it to feed primarily at the water surface, which is where it normally lives. Place plants around the sides of the tank so that, if stressed, the fish will not swim into the glass and damage its jaw. The male's anal fin is modified into a gonopodium. Breeding is more difficult than in the Poeciliidae, as the water must be very hard (add a 1 teaspoon (5 ml) of salt to 3 gallons (11 L) of water), and the fry need plants at the surface in which to hide. Fry feed on brine shrimp nauplii or fine dried food. At birth, the upper and lower jaws are almost the same length, but the lower jaw starts to grow longer after a few weeks.

3 BUTTERFLY GOODEID
Ameca splendens
Size: **3 in. (7.5 cm)**

This cousin of the red-tailed goodeid likes very hard or slightly brackish water.

Origin:	Number of fish per aquarium:	Diet:
Mexico	At least 1 pair	Dried food, small live and frozen foods

4 HALFBEAK
Dermogenys pusilla
Size: **2$\frac{1}{2}$–3 in. (6.5–7.5 cm)**

This is one of the few livebearers native to Asia (Hemirhamphidae family). It produces few offspring.

Origin:	Number of fish per aquarium:	Diet:
Indonesia, Thailand	At least 1 pair	Dried plant-based food, small live and frozen foods

A brackish-water aquarium

1 BUMBLEBEE FISH
Brachygobius xanthozonus

Family: Gobiidae
Size: 2 in. (5 cm)

Origin:	Number of fish per aquarium:	Diet:
Southeast Asia	5 or 6 per 15 gal. (55 L)	Small live food

2 MONO
Monodactylus argenteus

Family: Monodactylidae
Size: 6–8 in. (15–20 cm)

Origin:	Number of fish per aquarium:	Diet:
African and Asian coasts	4 to 6 per 50 gal. (190 L)	Omnivorous

3 BANDED ARCHERFISH
Toxotes jaculatrix

Family: Toxocidae
Size: 6–8 in. (15–20 cm)

Origin:	Number of fish per aquarium:	Diet:
Asia	1 pair	Live food at the surface or outside the water

4 GREEN PUFFERFISH
Tetraodon fluviatilis

Family: Tetraodontidae
Size: 6 in. (15 cm)

Origin:	Number of fish per aquarium:	Diet:
Southeast Asia	1 pair	Snails, worms, plants

5 SPOTTED GREEN PUFFERFISH
Tetraodon nigroviridis

Family: Tetraodontidae
Size: 6 in. (15 cm)

Origin:	Number of fish per aquarium:	Diet:
Southeast Asia	1 pair	Snails, worms, plants

6 SPOTTED SCAT
Scatophagus argus

Family: Scatophagidae
Size: 10–12 in. (25–30 cm)

Origin:	Number of fish per aquarium:	Diet:
Asia and Pacific islands	4 to 6 per 105 gal. (400 L)	Omnivorous, but predominantly plants

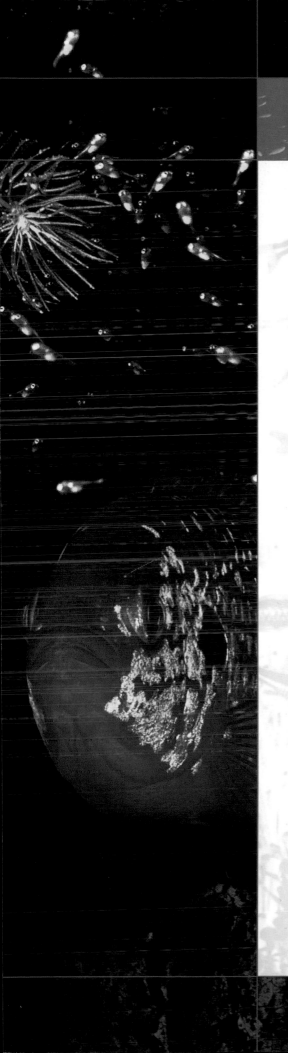

BREEDING
FRESHWATER FISH

Every aquarist yearns to breed fish and witness the birth of baby guppies or proudly revel in the sight of a shoal of fry growing up in the company of a pair of cichlids. Breeding aquarium fish can only be considered, however, if the prospective parents are provided the best possible environment. This is the secret of the professionals who breed freshwater fish on a large scale to supply aquarium stores. These experts select the most resistant strains and are able to develop new body shapes and colors in species such as fighting fish, guppies and platies.

On a less commercial level, many aquarists manage to breed species that are very fussy about their environment. In some cases, a simple change in diet can trigger spawning, but in others it is often necessary to reproduce water conditions that closely match those found in the fish's natural habitat. Then there is the challenge of raising the fry, which can sometimes be difficult to feed. However, successful feeding can usually be achieved by breeding suitably sized live food that is made available at appropriate times.

The breeding tank

Many aquarists do not breed their fish, as they think it is too complicated. It is typical, however, for a great many species to lay eggs in an aquarium, sometimes on a regular basis. The key is to provide conditions close to those existing in the wild.

This is especially true of freshwater species, as it is unusual to find saltwater fish that reproduce in captivity. An initial failure is disappointing and discouraging, but it is often due to a lack of knowledge. Successful breeding demands that certain preconditions are satisfied regarding the fish. Obviously, a male/female pair must be available, but it is not always easy to determine this.

Although some fish reproduce spontaneously in a community aquarium, it is advisable to use a separate tank for this purpose, so that the adults can spawn without being disturbed by other species and the fry can reach maturity in peace.

Tank size and setups

The breeding tank does not need to be very big, since it will only house the prospective parents (generally one pair, but sometimes more). It is not essential to provide a substrate, other than for a few species that lay their eggs on it. In some cases it may be necessary to insert fine mesh to separate the male from the female before or after the eggs are laid. In this small tank (and the small amount of water in it), fry will be able to find their food easily but, as they grow, and especially when there are many of them, you must think about transferring them to a larger aquarium.

Light

In some cases, strong lighting is the trigger for a fish to lay eggs, but other species prefer muted lighting or even almost total darkness. For these fish the lighting should be switched off and the sides of the breeding tank masked with dark paper.

Supports for the eggs

Some fish lay adhesive eggs on a support, such as plant foliage, rocks, wood or sometimes even the glass panels of the aquarium. Plants often used for this purpose include large clumps of Java moss *(Vesicularia dubyana)*, which protect the eggs from parents that may be inclined to eat them. You can also use artificial supports, particularly a homemade "spawning mop" (see Making a spawning mop,

A typical breeding setup

Fine-leaved plants such as Java moss hide the eggs from the parents.

Wood and rocks can provide spawning sites as well as shelter for some fish.

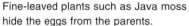

Breeding in acidic water

Java fern as a support for the eggs.

A waterlogged peat substrate will acidify the water.

Soft, acidic water (pH 6–6.5).

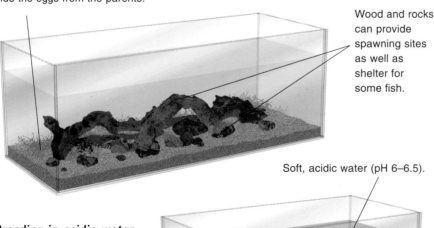

Breeding fish is ecologically sound

It aquarists make arrangements to trade, give away or sell fry to other hobbyists, this reduces the need to buy species collected in their original habitat, thereby helping to maintain their natural stock.

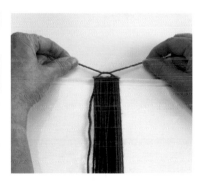

Above:
Rocks can serve as a support for the eggs.

Far left:
Some fish lay their eggs on or under the surface of a large leaf.

Left:
Eggs expelled in open water fall into the Java moss and are protected from predation by their parents.

opposite). It is held down by ballast or suspended from the surface of the aquarium attached to a cork. The mop resembles a water plant, and fish are willing to lay their eggs on it. It can be any size or thickness and placed exactly where the fish want to spawn. It is more practical than a plant, since it can be sterilized in hot water then reused, and moved from one aquarium to another.

Making a spawning mop

Wind nylon wool around a piece of cardboard about 30 times.

Slide an 8-inch (20 cm) piece of wool underneath and tie a knot.

Breeding traps

These are mainly used for breeding livebearers. The fry can escape predation by their mother, housed in one side of the unit, by swimming through the holes that give them access to the other side. The problem is that the breeding units sold in aquarium stores are often too small, and adult fish can injure themselves. Some may even abort their young. It is possible to make a reasonably sized breeding trap, but the most practical option is a small, densely planted rearing tank where fry can hide in the vegetation.

Cut the wool on the opposite side.

The mop is ready for use.

Water quality

Many species require water with certain characteristics – specifically particular levels of hardness and acidity. Many characins, for example, need very soft, acidic water. If the characteristics of the water required for reproduction are markedly different from those of the community tank, the prospective parents must be gradually adapted to their new surroundings in a breeding tank.

Above:
It is vital to maintain the hardness and pH values for species with precise requirements.

Right and below:
Peat can be placed in a tank to acidify the water. It becomes saturated with water and sinks to the bottom.

Hardness

Tap water is often too hard and needs to be softened. Mixing it with reverse-osmosis water, in specific proportions, is usually sufficient to solve this problem. It is also possible to collect rainwater and filter it through activated carbon. However rainwater contains all kinds of pollutants (particularly in industrialized areas) that can harm eggs, fry or even adult fish.

To increase the hardness of water, you can filter it through coral sand.

pH

It is sometimes essential to acidify water (and so reduce its pH) to enable certain fish to spawn. The most practical solution is to filter rainwater through activated carbon and then with peat until the desired pH is attained. You can also put a 2-inch (5 cm) layer of peat on the surface of a tank filled with rainwater. A week later, the peat will have sunk to the bottom and acidified the water, which will now resemble weak tea. The tank will then be ready for breeding fish native to South America.

Coral sand can be used to increase the hardness and pH.

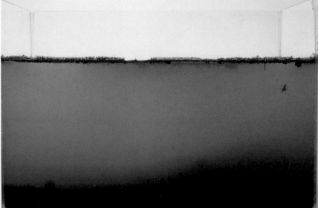

Temperature

Raising the temperature a few degrees relative to normal conditions sometimes stimulates fish to spawn and mature.

Filtration

This must be gentle, to avoid stressing the fish or sucking up the fry. In most cases, a small bubble-up sponge filter, with a fairly gentle flow, is sufficient for one pair of fish in a breeding tank.

Water changes

Even in a well-filtered breeding tank, partial and regular water changes (around 10 percent of the volume per week) will enhance both the health and growth of the fry. Obviously, the added water must have the same characteristics as those of the breeding tank. Gradually increase water changes to 50 percent twice weekly.

Above:
Some fish reproduce in neutral water with minimal hardness, as these are the characteristics of their natural habitat; this is the case with the krib (*Pelvicachromis pulcher*), native to Africa.

Left:
A small internal filter fitted with a simple block of foam as a filtering medium is sufficient in a breeding tank. The filtration will therefore be moderate, but the oxygenation will be effective.

Breeding stock

Except in the case of fish that reproduce spontaneously, prospective parents must be chosen with great care. You must avoid any fish that are deformed, sick or showing symptoms that suggest a pathological problem. Select the finest specimens, with well-developed fins (a sign of good health). Another factor is their coloring, which must be bright and bold.

Above:
It is not always easy to distinguish the sexes outside the breeding season. In the spawning season, the female's stomach swells up with eggs, and the male's coloring is often brighter, as seen in the upper of these two dwarf neon rainbowfish *(Melano-taenia preacox).*

Right:
In order to avoid ending up with deformed fry, it is important that the parents do not come from the same mother. In such cases, it is prudent to swap fish with another fishkeeper or buy more stock.

Consanguinity

It is important to avoid crossbreeding between blood relatives, i.e., between fish from the same mother. This could result in eggs that fail to hatch or fry that are deformed. Exchanging potential parents with other fishkeepers – through a club, for example – can help to overcome this difficulty.

Distinguishing the sexes

This is easy in those species in which one sex (normally the male) is more colorful than the other. There is no problem with livebearers, either, as the male can be recognized by its modified anal fin, known as a gonopodium. In other cases, there is often only a minimal difference between the two sexes – usually in the size and shape of the fins or the color.

It is almost impossible, however, to distinguish the sexes of most species outside the breeding period. It is only when one particular fish displays a swollen stomach for several days (not only after a good meal!) that it becomes apparent that this is a mature female. At this point, males can often be recognized by their behavior: courting display, defence of the spawning territory and driving away other fish.

Crossbreeding and hybrids

Crossbreeding with two parents of different colors often produces startling results! It is, however, wise to be wary of hybrids, i.e., crossbreeding between fish from different species but the same genus or family. It is possible that the fry may survive, with a mixture of their parents' traits, but they sometimes prove to be sterile. Although this phenomenon occurs spontaneously in

206

development of their off-spring. Also remember that laying eggs involves a significant outlay of energy on the part of a female, so this exertion must be compensated by providing high-quality nutrition.

Food should be served two or three times a day, and it should all be eaten within five minutes. Overfeeding obviously affects the water quality in the breeding tank.

Left:
Parents must be chosen chiefly on the basis of their coloring and clean bill of health, as shown in this male licorice gourami *(Parosphromenus deissneri)*. Listless or stressed fish will be less fertile.

Below:
The female's diet requires particular attention, as nutrients derived from her food will form the egg sac – essential during the fry's first days of life

the wild, it is not recommended in an aquarium. In fact, hybrids can be mistaken for an undiscovered species, which only adds to the confusion that already exists in the identification and classification of wild fish (this is particularly true with the cichlids of Lake Malawi).

Feeding the parents

This is probably the most important factor for successful breeding, but it is often over-looked.

Prospective parents need a balanced diet, if only to maintain them in good health. Also keep in mind that a nutritious diet plays an important role in ensuring viable eggs and egg sacs to sustain the future fry. Feeding parents properly therefore enhances the

Below:
Live foods for breeding stock. Left to right: water fleas *(Daphnia* spp.), bloodworm and adult brine shrimp.

A varied diet for breeding stock

The most important rule is to vary and alternate the food: provide good-quality dried food, but above all live and frozen foods (brine shrimp, worms), meaty foods (fish, mussels, shrimp) and algae (if they form part of the parents' dietary requirements).

Raising the fry

Although the fry of some species are protected and watched over by their parents, most are left to fend for themselves. In this case, the parents must be removed from the breeding tank as they will show no interest in their fry or may even try to eat them.

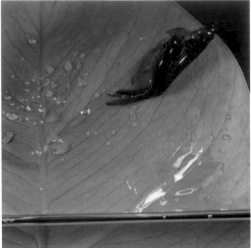

small mouth. Some fry are big enough to accept the fine powdered foods that are widely sold. Liquid fry foods are also available. Others must be given small live foods (whose movements will attract the fry). The most commonly used live foods are brine shrimp nauplii and infusorians, cultured by the fishkeeper.

Top right:
The male splash tetra *(Copella arnoldi)* ensures that the eggs stay moist by splashing them after they are laid on a leaf outside the water.

Above:
Sometimes, the fry are protected by their mother.

Right:
The egg sac is still visible on this recently hatched fry.

The egg sac
As soon as the fry emerge from their eggs or their mother, they start to use up the reserves in their egg sac. When these are almost exhausted after a day or so, the fry start looking for food, which must now be supplied by the fishkeeper.

How to feed the fry
Juveniles require larger portions than adults, as they grow quickly and need energy, primarily in the form of proteins. They should be served food three to five times a day, but must not be overfed. The main consideration is the size of the food, as this must be in proportion to the fry's

Culturing brine shrimp nauplii

Many species can take newly hatched brine shrimp as a first food, but it is always wise to feed some infusorians to start with along with the brine shrimp. That way, if the fry are not big enough to handle the larger food, they will not starve to death. Although specific equipment for culturing brine shrimp is available in aquarium stores, any container made of a food-safe material can be used for this purpose: a plastic bottle or bowl, or even a small tank. Dried brine shrimp eggs are available from many aquatic outlets, supplied in sealed cans or small glass jars. If you only use a small amount of eggs, store the remainder in a dry, cool place, otherwise they will fail to hatch after a few months.

Half-fill the culturing container with fresh tap water and $1^1/_2$ teaspoons (7.5 ml) of sea salt. Add $^1/_4$ teaspoon (1 ml) of eggs and drop in an air line attached to an air pump – the bubbles will circulate the eggs. After about 36 hours at 75°F (24°C) the eggs will have hatched and the air line should be removed. The shrimp will collect at the bottom and the shells will float to the surface. After a further 30 minutes, siphon out the shrimp with some air line and filter them through a paper towel. Wash them in freshwater and feed them to the fry.

Salt

Brine shrimp eggs

Culturing infusorians

Infusorian is a general term used for the many different microscopic organisms that live in water and feed on rotting plant matter. The spores are airborne, so no starter culture is required, only a suitable medium and food. Take an open jar of aquarium water and drop in a piece of slightly boiled potato. After about a week, the water will be cloudy with infusorians. To feed the fry, pour some of the cloudy water into the tank and top up the jar with fresh aquarium water. Having about five cultures available at any one time should keep the fry well supplied with infusorians.

Breeding cyprinids

Some of these fish are very easy to breed, even for a novice fishkeeper. Even better, they can lay eggs in a mixed tank, in the presence of other fish. The fry are robust and grow quickly.

Some cyprinids, such as rasboras, lay adhesive eggs on the fine leaves of plants, while others, like danios, expel eggs into open water so that they tumble down onto the substrate. A pair introduced into a breeding

Cabomba is an ideal fine-leaved plant to trap adhesive eggs.

Cherry barbs (*Puntius titteya*) spawning in a clump of fine-leaved plants. The male is the more colorful fish.

tank may produce eggs the following day, often triggered by sunlight striking the tank. Most parents are likely to eat their offspring, so they must be removed once the eggs are laid. The fry require soft, neutral or slightly acidic water, and they must initially be fed with infusorians.

Breeding zebra danios
(Danio rerio)

These egg-laying fish will readily breed, but the small size of the fry may cause problems for beginners. It is a good idea to cover the bottom of the breeding tank with a double layer of glass marbles. The eggs produced by the female will fall between the marbles and escape the attention of the parents, who must be removed immediately after the eggs are laid. Eggs hatch in three to four days and the fry will start swimming a few days later. Feed them first on infusorians or a liquid fry food and then with brine shrimp nauplii.

A pair of zebra danios spawning near the substrate. The male is on the left.

Breeding tank for zebra danios

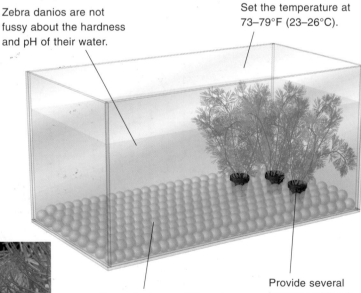

Zebra danios are not fussy about the hardness and pH of their water.

Set the temperature at 73–79°F (23–26°C).

Cover the base of the tank with large-grade gravel or a double layer of glass marbles to provide sanctuary for the eggs.

Provide several bunches of fine-leaved plants and some gentle aeration.

Breeding characins

Most characins kept in aquariums, such as tetras, are typical egg-scatterers, so their breeding tanks need plants or spawning mops for the females to lay their eggs on. In the wild, these fish spawn in deep shade, so keep the fry in low light.

Above: Warm soft water – home to many characins.

Most characins scatter their eggs into open water. They do not concern themselves with either the eggs or the fry. Tufts of Java fern (*Mycrosorium* spp.) or spawning mops will provide them with protection. They need soft, acidic water at 77°F (25°C). As the fry of most species are sensitive to light, keep the breeding tank dark after the eggs are laid.

Java fern is a good support for eggs.

Breeding neon tetras
(Paracheirodon innesi)

Despite a reputation for being difficult to breed, neon tetras will readily spawn in a community aquarium. Hatching and raising the fry proves more difficult. Condition the male and female separately and add them to a breeding tank with a peat substrate and clumps of Java moss. The water must be acidic (pH 6) and very soft. The eggs will be laid two or three days after the introduction of the parents. The male pursues the female, and the eggs fall out of reach within the fine foliage of the Java moss. Remove the parents after spawning. The eggs hatch within 24 hours and the fry feed on infusorians and brine shrimp nauplii. They acquire their brilliant coloring when they reach about ¹/₂ inch (1.25 cm) long.

Above:
A plump female neon tetra in breeding condition may produce 50–100 eggs per spawning every week.

Left:
A brightly colored neon tetra male in ideal breeding shape.

Breeding livebearers

The most popular live-bearing fish are platies, mollies, guppies and swordtails. These fish hold a special fascination, especially for beginners to fishkeeping, because they produce fully formed young. They are also highly prolific and will breed readily in a mixed community aquarium without any special attention.

A pair of hybrid sailfin mollies. The male is below, sporting a sail-like dorsal fin.

The breeding process in livebearers begins with a courtship display; then the male impregnates the female by moving his gonopodium (modified anal fin) toward the female's genital orifice, where he deposits his sperm. Most species can store sperm, so that more than one brood will be produced from a single mating. The fertilized eggs remain within the female and are released as fully formed young on a monthly basis. It is therefore vital to keep expectant mothers well fed at all times. Move females to a separate tank when they are about to drop their young (they become agitated when ready). Once the young are released they will swim away in search of food. Remove the mother at this stage and provide fry with fine foods such microworms, brine shrimp nauplii or finely powdered dried food.

Breeding tank suitable for mollies

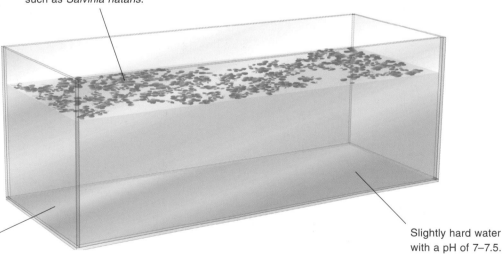

Only swollen, already impregnated females are put into the breeding tank.

Include some floating plants, such as *Salvinia natans*.

Set the temperature at 75°F (24°C).

Slightly hard water with a pH of 7–7.5.

Breeding labyrinth fish

The fish in this group have developed several distinctive techniques for laying eggs and sheltering the fry: some lay eggs in a bubblenest that floats on the surface of the water, while other fish incubate the eggs in their mouth. The bubble-nesting species are the ones most widely seen in the aquarium hobby.

Most members of this group deposit their eggs in a bubblenest built on the surface of the water. This nest plays a double role: it ensures protection for the emerging fry and provides them with oxygen (which is lacking in their native environment). Floating plants prove very useful for improving the construction and stability of this nest. For many labyrinth fish, the water quality necessary for breeding is not of critical importance, while for others it needs to be soft

Breeding tank for thick-lipped gourami

Temperature of 75–78°F (24–25°C).

Water quality is not critical for this species.

Cave for female to hide in if harassed by the male.

Floating plants such as riccia.

and acidic. The male builds the bubblenest and courts the female. He entices her to approach the nest and, after several spawning embraces, gathers the floating fertilized eggs into the nest. Remove the female at this point because she may be harassed by the male. Once the eggs hatch, remove the male to protect the fry.

Below:
The male betta patiently builds his bubblenest before encouraging the female to lay eggs.

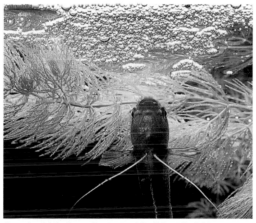

A male thick-lipped gourami builds a bubblenest.

Bettas
(Betta splendens)

The characteristics of the water are of little importance to this species. The male builds a bubblenest about 4 inches (10 cm) in diameter, before inviting the female to lay eggs. If she is unwilling, he may sometimes attack or even kill her. The fry feed on infusorians and brine shrimp nauplii. The male and female fry can only be distinguished from one another at about two months of age; they should then be raised separately.

A female betta swollen with eggs. She is ready to spawn.

Breeding tank for bettas

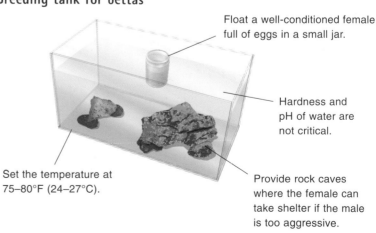

Float a well-conditioned female full of eggs in a small jar.

Hardness and pH of water are not critical.

Set the temperature at 75–80°F (24–27°C).

Provide rock caves where the female can take shelter if the male is too aggressive.

Breeding catfish

This is a huge group with a worldwide distribution and long history, and for some species, breeding in an aquarium is not an easy matter. The popular corydoras and suckermouth catfish, however, are among the least difficult to breed.

Right:
Bristlenose catfish (*Ancistrus* spp.) fry shortly after being born. Their egg sacs have not yet been exhausted.

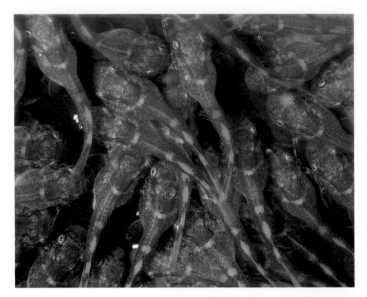

Catfish lay their eggs on a variety of supports: rocks, roots, long-leaved plants and even the tank glass. Small clay flowerpots are a practical alternative, as they are easy to move. For most species, the water should be soft and slightly acidic, at 77–82°F (25–28°C). Regular partial water changes are vital. For breeding catfish, it is best to use a tank dedicated to the task; eggs and fry may be lost in a community setup.

Peppered cory
Corydoras paleatus

In comparison to some other catfish, the peppered cory is easy to breed. The males have longer dorsal and pectoral fins; females appear very plump when they are ready to spawn. The male courts the female by brushing her with his barbels. When she is ready, the male curves his body to form an S-shape and releases sperm that the female takes up in her mouth. The sperm travels through her gut in a just a few minutes and is expelled through her vent over a batch of recently released eggs that are held between her pelvic fins. She then places the large, sticky fertilized eggs on a support (which she may clean first). Once the eggs hatch (after about four days), the parents should be removed. The fry are ready to feed after another three days. Provide brine shrimp nauplii and fine dried foods at this stage.

Above: A male suckermouth catfish guards the fertilized eggs.

Clay pots and bogwood are good spawning sites for catfish.

Breeding killifish

Some killifish – like those originally from surroundings that are always entirely submerged – lay their eggs on plants, but other species use highly original breeding strategies that reflect the yearly cycle of drought and rain in their native habitats.

There are two groups of killifish: those that lay eggs in plants and those that bury them in the substrate. The plant-spawners live in habitats where some water is available all year round. Their eggs hatch in about two weeks. The substrate-spawners live in shallow pools of water that evaporate in the dry season. To guarantee their survival, the eggs are buried in the soil and remain dormant until the first rains fall a few months later.

Killi lyretail
(Aphyosemion australe)

Unlike in many other killifish species, the eggs of the Cape Lopez lyretail do not undergo a diapause in their native habitat. To breed this fish, first prepare a small tank with a mop and a bed of peat as an alternative spawning medium. The pH must be 6.5, the hardness negligible and the temperature should be 73–75°F (23–24°C). Put one male with two to four females (the latter are distinguished by their rounded tail and drabber coloring). The spawning female produces about five (sometimes up to 20) eggs a day. Once the eggs are laid, remove the parents. The eggs hatch in 10–15 days. The fry can feed directly on brine shrimp nauplii; they become adults after three or four months.

The scientific term for this phenomenon is "diapause." This amazing natural cycle can be recreated in a small breeding tank with species such as *Notho-branchius eggersi*. Add a deep layer of peat to render the water acidic plus some plants – Java moss *(Vesicularia dubyana)*, for example – as a refuge for the females. Put one male together with two or three females for about a week. The spawning pair will deposit fertilized eggs in the peat; at this point all the adult fish should be removed. Next, siphon out the tank water and place the moist peat in a plastic container with small holes in the lid. (Aquarists who specialize in breeding these fish – "killiphiles" – can thus send their eggs by mail all over the world!) The eggs must remain damp for a few weeks (or several months for other species), before being returned to water, where they will hatch within a period of one to two weeks.

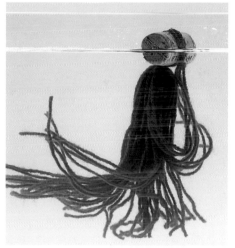

Top:
This killifish from Tanzania *(Notho-branchius eggersi)* is an annual fish: it dies when its surroundings dry up. It will have previously laid its eggs in the substrate. They hatch after a diapause of two to three months.

Above:
A spawning mop, used by some species as a support for eggs.

Breeding cichlids

Cichlids are a fascinating group of fish to breed because all provide some form of brood care for their eggs and fry. The two breeding strategies are laying eggs on the substrate and mouth-brooding.

Above:
A male jewel cichlid (*Hemichromis* spp., left) fertilizes the eggs that the female has just laid on a smooth rock.

Right:
Angelfish (*Pterophyllum* spp.) and discus (*Symphysodon aequifasciatus*) like to lay their eggs on the leaves of sturdy plants, such as this Amazon swordplant.

Center:
Other cichlids prefer to leave their eggs on a previously cleaned, flat rock.

Substrate spawners

South American cichlids prefer to lay their eggs on a rock, leaf, artificial support or directly on the aquarium glass. The water must be soft and acidic. One of the parents generally keeps watch over the eggs, fanning them to waft oxygenated water over them and removing any that rot. Guard duty includes repelling intruders.

Mouthbrooders

Mouth-brooding is a specialty of the cichlids that live in the African Rift Valley lakes, where the water is hard and alkaline. One of the parents – often the female – keeps the eggs in its mouth during the incubation period. Once the fry are hatched, they sometimes return to this shelter when danger threatens, until they are big enough to brave it alone. This protection ensures the maximum survival rate for young fish. The parent that looks after the eggs is unable to eat during the incubation, so it is essential to feed it well afterward.

A very maternal female

The male zebra mbuna (*Metriaclima zebra*) is polygamous and may breed with several females. It can be recognized by the large yellow patches on its anal fin. The female first keeps the eggs (30 on average) and then the fry in her mouth until the latter can swim freely, which takes a few weeks. She continues to take care of the fry for about another week. It is best to move a brooding female to a separate tank before she releases the fry. Once free-swimming, feed the fry on brine shrimp nauplii.

These discus fry, just a few days old, feed on the mucus on their parents' sides before going on to take small live food.

Parents that feed their fry

Given the right conditions, discus (*Symphysodon* spp.) are relatively easy to breed. They need warm water (82°F/28°C) that is very soft and slightly acidic (pH 6.5). The eggs take three days to hatch and the fry start swimming on the seventh day. Then they cluster around the flanks of their parents for about 10 days, feeding on the nutritious mucus secreted from their skin. After this period they will accept brine shrimp nauplii.

Breeding angelfish
(*Pterophyllum scalare*)

The parents clean a site on a vertical surface for the eggs. The female lays her eggs in rows, and the male then fertilizes them. The parents generally aerate the eggs, but it is possible that they may lose interest, particularly if it is their first spawning. If this happens, remove the support with the eggs and place it in a small tank, then gently stir the water in the vicinity with a stream of fine bubbles from an air line. The eggs will hatch in three days. Angelfish pairs remain faithful and lay eggs regularly.

Above and left:
Angelfish (*Pterophyllum scalare*) often lay their eggs on a vertical rock. The female is distinguished by the slight swelling of her belly behind the pelvic fins.

SALTWATER FISH

For both novice and experienced fish-keeper, there is nothing quite so beautiful as a saltwater fish! The diversity of their colors is only rivaled by that of birds. These colors are the key to their survival, for in the wild, all that glitters is poison.

The wide variety of their forms and behavior is fascinating for hobbyists, who can indulge in fantasies of deep-sea diving as they observe their saltwater fish. Unfortunately, saltwater species reproduce only very rarely in an aquarium. Although there are some specialized breeders entirely devoted to the task, it is more usual to take saltwater fish from their natural habitat. For some time now, however, a new capturing method has been used, which involves trapping the larvae of reef fish when they enter lagoons. After growing for a few weeks, the fry are chosen according to their suitability for an aquarium. Some are put back into the sea, while the others are left to grow in a tank.

Saltwater aquariums have probably safeguarded their future with these innovative techniques. They do have a negative side, but the arguments against them are not fully convincing.

Damselfish and clownfish
(Pomacentridae family)

Aquarists making their first foray into a saltwater tank generally start off with fish from this family, as these robust species are readily available and some may reproduce in an aquarium (unusual among saltwater fish). As the fish are a modest size, a 50-gallon (190 L) tank is sufficient for raising some species while still meeting their compatibility requirements.

1 YELLOWTAIL DAMSEL
Chrysiptera parasema
Size: **3 in. (7.5 cm)**

It is often confused with a related species, *C. hemicyanea* (azure damsel).

Origin:	Number of fish per aquarium:	Diet:
From the Philippines to New Guinea	2 or 3	Small live and frozen foods, dried food

Yellowtail damsel

This fish defends its territory energetically against both members of its own species and other damsels of a similar color. For this reason, it should be kept in at least an 80-gallon (300 L) tank with rocky decor. It does not pay any attention to invertebrates, however, and so can live alongside them.

1

2

2 BLUE DEVIL DAMSEL (SAPPHIRE DEVIL)
Chrysiptera cyanea
Size: **4 in. (10 cm)**

This fish quickly adapts to captivity and accepts a wide range of food.

Origin:	Number of fish per aquarium:	Diet:
Indian and Pacific Oceans, Philippines and Australia	1	Small live and frozen foods, dried food

Breeding damsels

Damsels are territorial fish. The male attracts the female to the spawning area but fends off other fish. The female lays her eggs on a hard surface, such as a rock or coral; the male fertilizes the eggs and aerates them with his fins. The eggs hatch in a few days, and the fry must then be fed very small live prey (even smaller than brine shrimp nauplii), such as rotifers. This saltwater zooplankton can be cultured by aquarists, but is often available through aquarium clubs.

Blue devil damsel (sapphire devil)

This damsel generally lives alone, as it is belligerent toward members of its own species. It does best in a living reef aquarium with open space for swimming. This fish is easy to rear, but should be kept away from invertebrates, as it likes to eat them.

Domino damsel

This fish must be given plenty of hideaways in the aquarium rockscape, as well as an open area suitable for swimming. It can live in a group when juvenile (as pictured here), but can become aggressive in adulthood. However, it proves easy to acclimatize and does not attack invertebrates. It can be bred in a tank (the male keeps watch over the eggs during incubation). The fry are difficult to feed. Another closely related species, *Dascyllus aruanus*, the three-stripe damsel, is less aggressive.

3

3 DOMINO DAMSEL
Dascyllus trimaculatus
Size: **3 in. (7.5 cm)**

Although the juveniles are fairly placid, the adults are aggressive.

Origin:	Number of fish per aquarium:	Diet:
Red Sea and parts of the Indian and Pacific oceans	1	Small live and frozen foods, dried food

Symbiosis

Clownfish take refuge from their enemies in the tentacles of an anemone and lay their eggs at its base, so they must be provided with an anemone in the aquarium. (The mucus on the clownfish's skin protects it from the venom secreted by the anemone's tentacles.) Anemones also benefit from the presence of clownfish, as they dispose of bits of leftover food. Some scientists call this mutually beneficial association "symbiosis," but others prefer the term "commensalism," as the two animals are not strictly interdependent.

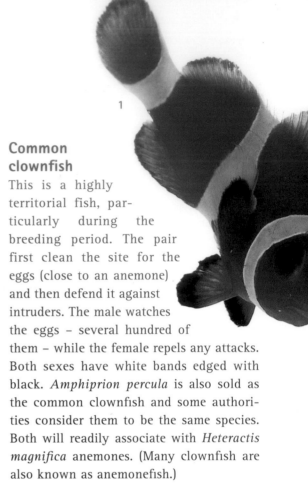

1 COMMON CLOWNFISH (WESTERN CLOWNFISH)
Amphiprion ocellaris

Size: 3 in. (7.5 cm)

One of the most common clownfish – it is easy to breed and feed.

Origin:	Number of fish per aquarium:	Diet:
Philippines, Japan, China Sea	2 or 3 (Associated anemones: *Heteractis magnifica, Stichodactyla gigantea*)	Small live and frozen foods, dried food

2 CLARKII CLOWNFISH (CLARK'S CLOWNFISH)
Amphiprion clarkii

Size: 5 in. (12.5 cm)

This is less aggressive than other clownfish.

Origin:	Number of fish per aquarium:	Diet:
Indian and Pacific Oceans	2 or 3 (Associated anemones: All those mentioned for other clownfish)	Small live and frozen foods, dried food

Clarkii clownfish (Clark's clownfish)

This fish usually lives in pairs and is not fussy with regard to the anemone species associated with it. Reproduction is easy as the pair defend their territory with less ferocity than other clownfish. The eggs hatch after a week, and the emerging fry are slightly bigger than those of other related species. They must be provided with small live foods, however, and this is the most awkward aspect of their breeding process, although it is not impossibly difficult.

Common clownfish

This is a highly territorial fish, particularly during the breeding period. The pair first clean the site for the eggs (close to an anemone) and then defend it against intruders. The male watches the eggs – several hundred of them – while the female repels any attacks. Both sexes have white bands edged with black. *Amphiprion percula* is also sold as the common clownfish and some authorities consider them to be the same species. Both will readily associate with *Heteractis magnifica* anemones. (Many clownfish are also known as anemonefish.)

Tomato clownfish

This species is not difficult to breed, although the normally placid parents will become extremely aggressive and chase away all intruders. They completely ignore invertebrates, however.

Pink skunk clownfish

It has a white horizontal band on its back and another vertical one on its operculum. The band on the operculum is absent in a related species, the orange skunk clownfish (*Amphiprion sandaracinos*), which also lives with the anemone *Heteractis magnifica*. Both these species live in pairs and are easy to keep in captivity.

Other clownfish

A. bicinctus (two-band clownfish): only aggressive toward members of its own species, easy to acclimatize. Associated anemones: *Entacmaea* spp., *Heteractis* spp.

A. ephippium (saddle clownfish): one of the largest clownfish. Associated anemones: *Entacmaea* spp., *Heteractis* spp., *Stichodactyla* spp., but it can do without one.

Premnas biaculeatus (maroon clownfish, spinecheek anemonefish): it has a spine just behind each eye. It is probably the most aggressive of all. Associated anemone: *Entacmaea* spp., but it can do without one.

3 TOMATO CLOWNFISH
Amphiprion frenatus

Size: **5 in. (12.5 cm)**

This is robust and easy to breed.

Origin:	Number of fish per aquarium:	Diet:
Pacific and Indian oceans	**2 or 3** (Associated anemones: *Entacmaea quadricolor*)	**Small live and frozen foods, dried food**

4 PINK SKUNK CLOWNFISH
Amphiprion perideraion

Size: **4 in. (10 cm)**

The male's dorsal and anal fins are edged with a thin orange band.

Origin:	Number of fish per aquarium:	Diet:
Area between Thailand and New Caledonia	**1 pair** (Associated anemones: *Heteractis magnifica*)	**Small live and frozen foods, dried food**

A clownfish aquarium

1 ORANGE SKUNK CLOWNFISH
Amphiprion sandaracinos
Size: 4¹/₂–5 in. (11–12.5 cm)

Origin:	Number of fish per aquarium:	Diet:
Pacific Ocean	1 pair with an anemone	Omnivorous; live food

2 PINK SKUNK CLOWNFISH
Amphiprion perideraion
Size: 4 in. (10 cm)

Origin:	Number of fish per aquarium:	Diet:
Pacific Ocean	1 pair with an anemone	Omnivorous; live food

3 SADDLE CLOWNFISH
Amphiprion ephippium
Size: 4¹/₂–5 in. (11–12.5 cm)

Origin:	Number of fish per aquarium:	Diet:
Indonesian coasts	1 pair with an anemone	Omnivorous; live food

4 TOMATO CLOWNFISH
Amphiprion frenatus
Size: 4–4¹/₂ in. (10–11 cm)

Origin:	Number of fish per aquarium:	Diet:
Thailand, Indonesia, Philippines	1 pair with an anemone	Omnivorous; live food

5 TWO-BAND CLOWNFISH
Amphiprion bicinctus
Size: 4¹/₂–5 in. (11–12.5 cm)

Origin:	Number of fish per aquarium:	Diet:
Red Sea, Gulf of Aden	1 pair with an anemone	Omnivorous; live food

6 CLARKII CLOWNFISH
Amphiprion clarkii
Size: 4–4¹/₂ in. (10–11 cm)

Origin:	Number of fish per aquarium:	Diet:
Indian and Pacific oceans	1 pair with an anemone	Omnivorous; live food

Angelfish (Pomacanthidae family)

The common characteristic of these fish is the spine on each of their gill covers. This family includes both modestly sized fish and species worthy of a place in a public aquarium. The former include the dwarf, or pygmy, angelfish (from the *Centropyge* genus), which barely exceed 5 inches (12.5 cm) in length. The latter boast some of the most majestic of all aquarium fish. The juveniles' coloring is often very different from that of the adults, often leading to confusion over their names. Some species are fussy feeders and can be difficult to acclimatize. No species from this family has been known to reproduce in an aquarium.

1 BICOLOR ANGELFISH
Centropyge bicolor

Size: **5 in. (12.5 cm)**

One of the most common angelfish; its coloring is unmistakable.

Origin:	Number of fish per aquarium:	Diet:
Area between Malaysia, Japan and northwest Australia	1	Small live and frozen foods, dried food

Bicolor angelfish

This is a peaceful, even timid dwarf species when faced with other fish, but it can be aggressive toward members of its own species. It will thrive in a minimum 50-gallon (190 L) aquarium furnished with plenty of hideaways. A related species, *Centropyge argi* (cherub angelfish), found in the Caribbean, is smaller, and only the front part of its body is yellow.

2

2 FLAME ANGELFISH
Centropyge loricula

Size: 4 in. (10 cm)

This fish owes its name to its vivid red coloring.

Origin:	Number of fish per aquarium:	Diet:
Philippines, Marshall Islands, Hawaii	1	Small live and frozen food, dried food

3 CORAL BEAUTY ANGELFISH
Centropyge bispinosa

Size: 5 in. (12.5 cm)

This fish cohabits placidly with fish of the same size, apart from those of its own species.

Origin:	Number of fish per aquarium:	Diet:
South Africa, Indian and Pacific Oceans	1	Small live and frozen food, dried food

Other dwarf angelfish

C. acanthops (yellow and blue angelfish): yellow with a blue ventral patch behind the operculum. Rarely available commercially.

C. eibli (Eibl's angelfish, red-stripe angelfish): gray, with a black upper body. It must be acclimatized with algae as its first food. Fairly aggressive toward species from the same genus, but easy to keep.

C. ferrugata (rusty angelfish): reddish brown color with dark patches.

C. flavissima (lemonpeel angelfish): completely yellow, apart from a blue ring around the eyes and a blue patch on the gill cover.

C. heraldi (yellow angelfish): similar to *C. flavissima*, but entirely yellow, with no blue markings.

C. potteri (Potter's angelfish): similar to the above, but with thin blue stripes on the dorsal and anal fins and a blue patch on the body to the caudal fin.

C. vrolikii (pearlscale angelfish, half-back angelfish): a little bigger than those above, easy to acclimatize, but less readily available.

Flame angel

The flame angelfish treats all other fish with equanimity, apart from those of its own species. It appreciates hiding places between blocks of coral and water that is well aerated and filtered. It will graze on the green algae that may sometimes appear on the rockscape.

Coral beauty angelfish

This is a hardy fish that is easy to rear and generally safe with invertebrates. Adult coloring varies, but the dark vertical bands are always very pronounced.

3

Emperor angelfish

This fish is highly territorial and behaves aggressively toward other angelfish, although it is relatively sociable with other species. It requires a tank of at least 105 gallons (400 L), furnished with hiding places and filed with well-filtered and aerated water. Depending on its origin, its dorsal fin is round or pointed, and its caudal fin yellow or orange. The juveniles are dark blue with white, sometimes circular, marks. Juveniles adapt to captivity more easily than adults.

1 EMPEROR ANGELFISH
Pomacanthus imperator

Size: **12 in. (30 cm)**

This is one of the most beautiful and majestic of all angelfish.

Origin:	Number of fish per aquarium:	Diet:
Red Sea, Indian and Pacific Oceans	1	Small live and frozen foods, plants, dried food

2 KORAN ANGELFISH
Pomacanthus semicirculatus

Size: **10 in. (25 cm)**

Neither juveniles nor adults tolerate members of their own species.

Origin:	Number of fish per aquarium:	Diet:
Red Sea, Indian and Pacific Oceans	1	Small live and frozen foods (particularly worms and mussels), dried food

Koran angelfish

It is easier to acclimatize the juveniles as they willingly accept live food. Their dark blue bodies are marked with broken, white circular bands (setting them apart from *Pomacanthus imperator*). The adults are dark brown; the white bands gradually disappear over time. They do best in an aquarium with plenty of nooks and crannies so they can establish their territory. These fish leave their shelter to eat.

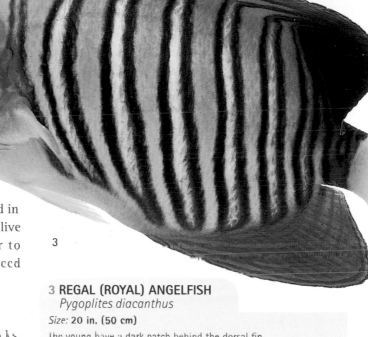

Regal (royal) angelfish

This fish cannot bear the presence of other angelfish in its territory. However, it is sociable toward other species. It is difficult to acclimatize, as its diet in the wild is based on sponges. Patience is therefore required in offering a range of substitutes, especially live brine shrimp. The young are easier to acclimatize, but they are timid and need plenty of hideaways.

Queen angelfish

An aquarium furnished with coral or rocks, complete with hideaways, helps this fish to adapt and enables it to mark out its territory. Although it is aggressive toward other angelfish, it respects other species. The juveniles acclimatize fairly easily and enjoy a varied diet; they particularly appreciate live food and algae. Juveniles sport three bluish vertical bands on their flanks and two more around the eye, which disappear when they reach maturity. A closely related species, *Holacanthus tricolor* (rock beauty angelfish), is more difficult to keep in an aquarium: it is vital to introduce them into the tank while they are still juveniles.

3

3 REGAL (ROYAL) ANGELFISH
Pygoplites diacanthus
Size: 20 in. (50 cm)
The young have a dark patch behind the dorsal fin.

Origin:	Number of fish per aquarium:	Diet:
Red Sea, Indian Ocean	1	Sponges, small live and frozen food, worms

4

4 QUEEN ANGELFISH
Holacanthus ciliaris
Size: 12 in. (30 cm)
This fish lives alone and is aggressive toward other angelfish.

Origin:	Number of fish per aquarium:	Diet:
Caribbean, tropical Atlantic Ocean	1	Small live and frozen foods, mussels, worms, plants

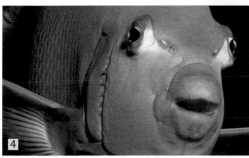

4

Butterflyfish (Chaetodontidae family)

Butterflyfish resemble angelfish but they lack the spine on the gill cover and have a pointed head. This group includes some of the most splendid species found in coral reefs. These fish are often fussy feeders in an aquarium.

1

1 THREADFIN BUTTERFLYFISH
Chaetodon auriga
Size: 8 in. (20 cm)

This fish is characterized by the adult's filamentous extension on the dorsal fin.

Origin:	Number of fish per aquarium:	Diet:
Red Sea, Pacific Ocean	1	Coral polyps, algae, small live and frozen foods

Raccoon butterflyfish

This fish does not tolerate members of its own species and can even react aggressively to other fish if the aquarium is too small. It requires hiding places and open

Threadfin butterflyfish

This reef fish is only aggressive toward members of the same species or butterflyfish with similar coloration. As it mainly feeds on polyps, it must not be kept with live corals. It likes hiding places among the aquarium decor and enjoys eating brine shrimp.

A related species, the vagabond butterflyfish (*Chaetodon vagabundus*), can be distinguished by a black band instead of a patch toward the rear of the body.

2

2 RACCOON BUTTERFLYFISH
Chaetodon lunula
Size: 8 in. (20 cm)

This species is easy to breed, as long as it is kept away from invertebrates.

Origin:	Number of fish per aquarium:	Diet:
Red Sea, Indo-Pacific	1	Small live and frozen foods, algae, dried food

space where it can swim freely. The juveniles sport a black patch on the caudal peduncle; this turns into a band in adults. It is closely related to *Chaetodon fasciatus* (diagonal butterflyfish), which is more difficult to keep in captivity.

Other butterflyfish

Other species from the *Chactodon* genus are sometimes stocked by aquarium dealers, particularly *C. vagabundus* (vagabond butterflyfish) and *C. collare* (red-tailed butterflyfish), both of which are easy to rear.

3 COPPERBAND BUTTERFLY-FISH *Chelmon rostratus*

Size: 6 in. (15 cm)

Its large "beak" enables this fish to capture its prey in coral reefs.

Origin:	Number of fish per aquarium:	Diet:
Indo-Pacific	1	Small live and frozen foods, especially brine shrimp and mussels

Copperband butterflyfish

This species is easy to acclimatize, especially when it is young, but it does not tolerate fish of its own species. The black patch on the dorsal fin resembles an eye and acts as a deterrent to its enemies.

Longnose butterflyfish

This fish lives alone and is very sociable toward other fish, apart from those of its own species. It is not really aggressive: it restricts itself to flaunting its dorsal fin to make an impression. It is sensitive to the quality of the water, which must be well filtered and partially changed on a regular basis. A closely related species, *Forcipiger flavissimus*, strongly resembles it but has a shorter "nose."

4 LONGNOSE BUTTERFLYFISH *Forcipiger longirostris*

Size: 8 in. (20 cm)

Its oversized "nose" allows it to seek out food in every nook and cranny.

Origin:	Number of fish per aquarium:	Diet:
Hawaii, East Africa, Indo-Pacific	1	Small live and frozen foods, especially crustaceans and mussels

Groupers and basslets
(Serranidae, Grammatidae and Pseudochromidae families)

The Serranidae family includes the snappers; only the smallest of these are kept in aquariums. They are carnivores and do not hesitate to eat small fish. There are a few small species in related families that are suitable for an aquarium, and they enhance it with their bright colors.

Panther grouper
This species is sociable with all other fish of the same size, but smaller ones may be regarded as prey by this fairly voracious carnivore. The patches are darker and more marked in the juveniles.

Coral hind
This grouper calmly ignores other fish of the same size, but it cannot tolerate its own species. However, it does take notice of small fish – to eat them. The adults are darker than the juveniles.

Sea goldie (lyretail anthias)

This fish likes to live in a small group and therefore requires a fairly large tank. It is sociable toward all other fish and takes no notice of invertebrates. The male is a mauve color and the third stripe on its dorsal fin is longer than that of the female, which is a golden-orange color.

Royal gramma

This fish's behavior is sociable, although it will not tolerate fish of its own species in a small tank. It is sometimes confused with *Pseudochromis paccagncllae* (royal dotty-back – Pseudochromidae family), which is native to the Indo-Pacific and nicknamed the false gramma. Both these species are ideal for an invertebrate tank as they will take no notice of their fellow residents.

3

4

1 PANTHER GROUPER
Cromileptes altivelis

Size: **20 in. (50 cm)**

It requires a large tank so that it can swim in all its glory.

Origin:	Number of fish per aquarium:	Diet:
Indo-Pacific, Australia	1	Small live and frozen food, small fish, dried food

2 CORAL HIND
Cephalopholis miniata

Size: **12 in. (30 cm)**

This is a solitary fish that can live for a relatively long time.

Origin:	Number of fish per aquarium:	Diet:
Indo-Pacific	1	Small live and frozen food, mussels, small fish, dried food

3 SEA GOLDIE (LYRETAIL ANTHIAS)
Pseudanthias squamipinnis

Size: **6 in. (15 cm)**

This species is from the same family as the snappers, but is markedly less voracious.

Origin:	Number of fish per aquarium:	Diet:
Red Sea, Indo-Pacific	1 group of variable size, depending on tank dimensions	Small live and frozen foods, dried food

4 ROYAL GRAMMA
Gramma loreto

Size: **3 in. (7.5 cm)**

This is a peaceful species that is easy to acclimatize and very popular on account of its bright coloring.

Origin:	Number of fish per aquarium:	Diet:
Caribbean	1, or more if the tank is big enough	Small live and frozen food, mussels, fish, dried food

Wrasses (Labridae family)

These species owe their name to their thick lips (*labra* in Latin). They prove very attractive in an aquarium, as their colors are often bright, and there are marked differences between the juveniles and the adults. Some species bury themselves in the substrate or make a cocoon that serves as a "sleeping bag." There is currently no record of their reproduction in an aquarium.

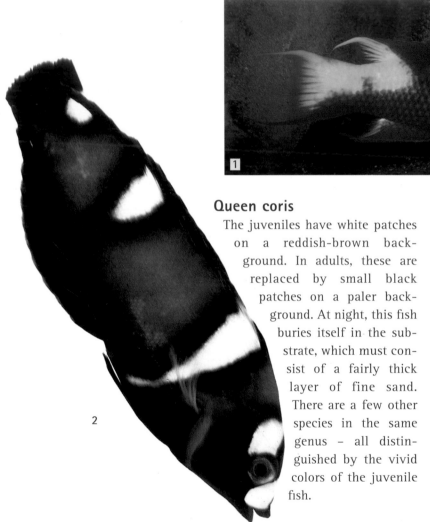

Queen coris

The juveniles have white patches on a reddish-brown background. In adults, these are replaced by small black patches on a paler background. At night, this fish buries itself in the substrate, which must consist of a fairly thick layer of fine sand. There are a few other species in the same genus – all distinguished by the vivid colors of the juvenile fish.

Spotfin hogfish

The Cuban hogfish is normally peaceful, except with other fish of the same species. It must not be kept with invertebrates as it may find them too appetizing to overlook.

Green birdmouth wrasse

This is a robust species that is easy to rear. Although its restless activity can upset more placid fish, it is peaceful. Its long "muzzle" allows it to retrieve food from crevices in coral, but it must not be kept with invertebrates. The adult male has a blue-green body, while the female and juveniles are a more brownish color. It does not burrow into the substrate at night, but hides among the coral.

1 SPOTFIN HOGFISH (CUBAN HOGFISH)
Bodianus pulchellus
Size: **6 in. (15 cm)**
There are several species of *Labridae* in the *Bodianus* genus, all with very vivid coloring.

Origin:	Number of fish per aquarium:	Diet:
Caribbean	1	Small live and frozen foods, dried food

2 QUEEN CORIS
Coris formosa
Size: **10 in. (25 cm)**
This wrasse does not tolerate fish of its own species, and it enjoys eating invertebrates.

Origin:	Number of fish per aquarium:	Diet:
Indo-Pacific	1	Small live and frozen foods, dried food

Harlequin tuskfish

The patches on the juveniles' fins are absent in the otherwise stunningly colored adults. Adult fish cannot bear each other's presence; they will fight using their strong teeth and are capable of causing serious injury. They are generally peaceful toward other species, although they may attack them on occasions, particularly at feeding time. This fish likes eating invertebrates, so do not keep them together.

3 GREEN BIRDMOUTH WRASSE
Gomphosus caeruleus
Size: **10 in. (25 cm)**
Pair can coexist, but two males will not tolerate each other.

Origin:	Number of fish per aquarium:	Diet:
Pacific Ocean, Red Sea	1 or 1 pair	Dried food, small live and frozen foods

4 HARLEQUIN TUSKFISH
Choerodon fasciatus
Size: **8 in. (20 cm)**
This fish can dislodge pieces of the decor with its powerful teeth.

Origin:	Number of fish per aquarium:	Diet:
Pacific Ocean, from Japan to Australia	1	Small live and frozen foods, worms, mussels, dried food

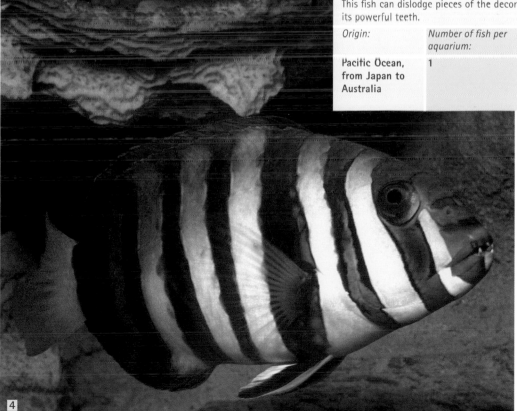

1 YELLOW WRASSE
Halichoeres chrysus
Size: **5 in. (12.5 cm)**
It can sometimes skirmish with other members of its species.

Origin:	Number of fish per aquarium:	Diet:
Pacific Ocean	2 or 3	Small live and frozen foods, dried food

2

the night. Another fish, the false cleaner wrasse *(Aspidontus taeniatus)*, takes advantage of its uncanny resemblance to the genuine cleaner wrasse to harass other fish and rip off their scales or gills.

Yellow wrasse

This is an ideal wrasse for an aquarium, as it is not very big, even when adult sized. The juveniles live in a group, but adults are solitary. Like all members of this family, it relishes invertebrates and cannot be kept with them.

Blue-streak cleaner wrasse

This fish's behavior is unusual. It writhes around in particular areas of the tank to attract the attention of other fish; it then removes the external parasites on the skin – and even on the gills by sliding under the gill cover. Some fish are aware of this behavior and specifically approach this wrasse to be cleaned. In the evening it goes into a hiding place and secretes a mucous cocoon, which serves as its shelter for

2 BLUE-STREAK CLEANER WRASSE
Labroides dimidiatus
Size: **5¹/₂ in. (14 cm)**
This fish is quite difficult to acclimatize. It must first be given live foods.

Origin:	Number of fish per aquarium:	Diet:
Red Sea, Indo-Pacific	1	Small live and frozen foods, dried food

Lionfish (Scorpaenidae family)

These fish project a deceptive image of serenity. In fact, they are capable of quickly gobbling up any prey that strays within their reach. Moreover, the spiny rays on their fins constitute a real danger, as they secrete venom. Lionfish, also called turkeyfish, must not be touched with bare hands, therefore, and you should see a doctor immediately in the event of an accident.

Red lionfish

The lionfish is given a wide berth by fish of its own size, and even larger ones. It needs a large tank furnished with hiding places. There are two closely related species that share the same characteristics; one of these, the clearfin lionfish (*Pterois radiata*), is slightly smaller.

Zebra lionfish

The membrane linking the rays on the pectoral fins is more developed than in the *Pterois* genus, which gives this fish an even more majestic appearance. It is not aggressive either toward fish of its own species or other lionfish.

3 RED LIONFISH
Pterois volitans
Size: **14 in. (35 cm)**

This is the most common lionfish; it chases away all other fish.

Origin:	Number of fish per aquarium:	Diet:
Indo-Pacific	1	Small live and frozen foods, fish, mussels, dried food

4 ZEBRA LIONFISH
Dendrochirus zebra
Size: **7 in. (17.5 cm)**

This cousin of the lionfish is a more modest size but shares the same behavior patterns.

Origin:	Number of fish per aquarium:	Diet:
Indo-Pacific	1	Small live and frozen foods, fish, mussels, dried food

Tangs (Acanthuridae family)

These are also known as surgeonfish, due to a small, scalpel-shaped spine on the caudal peduncle, which they can extend to defend themselves. They generally do not tolerate fish of their own species or of a similar color. Their diet is predominantly plant-based.

Powder blue tang

This tang's yellow spine is clearly visible, as is the white patch under its head. It is aggressive, territorial and will not tolerate the presence of another tang. A juvenile is easier to acclimatize in an aquarium – just give it plant-based food, in well-filtered and aerated water with tank decor that offers plenty of hideaways.

1

2

Japan surgeonfish (powder brown surgeonfish)

This fish is easily identified from above by the white patch between its eye and its mouth. Similar to other tangs, it must be the only member of its family in an aquarium, which should contain at least 80 gallons (300 L) of water and have plenty of hiding places. It is a strong swimmer that spends much of its time grazing the green algae on the decor. It can be difficult to acclimatize, as it sometimes refuses to eat.

1 POWDER BLUE TANG
Acanthurus leucosternon
Size: **10 in. (25 cm)**
Its small mouth enables it to efficiently graze on algae.

Origin:	Number of fish per aquarium:	Diet:
Indian Ocean	1	Algae, small live and frozen foods, dried food that is predominantly plant-based

2 JAPAN SURGEONFISH (POWDER BROWN SURGEONFISH)
Acanthurus japonicus
Size: **8 in. (20 cm)**
This species is more complicated to rear than the other tangs.

Origin:	Number of fish per aquarium:	Diet:
Indo-Pacific	1	Algae, small live and frozen foods, dried food that is predominantly plant-based

Lined surgeonfish

This species is easy to raise in a large tank furnished with hiding places, which it will weave in and out of effortlessly. It appreciates well filtered, moving water. This tang is boisterous and often chases the aquarium's other occupants, even if they are from another species.

There is another species of tang with horizontal stripes, native to the Red Sea, called the sohal tang *(Acanthurus sohal)*. It is bigger but exhibits the same behavior, albeit with slightly more aggression. In both these species, acclimatization is easier to achieve with juveniles.

Blue tang (regal tang)

Although the young can live in a group, the adults are solitary. This territorial fish needs hiding places to withdraw into at night. By day, it gently swims around in search of food, especially algae to graze on. The coloring can vary according to where a specimen was captured – and even from one fish to another.

3 LINED SURGEONFISH
Acanthurus lineatus
Size: **8 in. (20 cm)**
This tang requires plenty of space to feel at ease.

Origin:	Number of fish per aquarium:	Diet:
Indo-Pacific	1	Algae, small live and frozen foods, dried food that is predominantly plant-based

4 BLUE TANG (REGAL TANG)
Paracanthurus hepatus
Size: **8 in. (20 cm)**
This tang will tolerate other fish, even those from its own family.

Origin:	Number of fish per aquarium:	Diet:
Indo-Pacific	1	Algae, small live and frozen foods, dried food that is predominantly plant-based

1

Other tangs

Listed below are two other species of tangs that are available from aquarium dealers. They are aggressive toward members of their own species, but the juveniles are easier to acclimatize. Their diet must include algae or other vegetable matter.

The Achilles tang (*Acanthurus achilles*): The adults sport a red patch near the spine. Their water must be thoroughly aerated and highly oxygenated.

Orange-shoulder tang (*A. olivaceus*): It owes its name to the orange area behind its eyes. The juveniles are completely yellow.

Naso tang

As this fish is almost continuously on the move, it needs a large tank with plenty of open space to swim freely. It is less aggressive toward its own species than other tangs. The juveniles' dorsal fin is black, while that of adults is yellow.

1 NASO TANG
Naso lituratus

Size: **12 in. (30 cm)**

The spine on its caudal peduncle is permanently raised.

Origin:	Number of fish per aquarium:	Diet:
Red Sea, Pacific Ocean	1	Algae, small live and frozen foods, dried food that is predominantly plant-based

2

Yellow tang

This tang is sociable and relatively tolerant of fish of its own species: it does not attack them, but does chase them out of its territory. Its diet makes it a perfect aquarium resident, as it clears up unwanted algae that would otherwise be difficult to eliminate. As it swims practically all the time, it needs plenty of free space.

Desjardin's sailfin tang

Young fish (less than 4 inches/10 cm long) are easier to acclimatize than adults; they should be fed vegetable matter or animal foods. Juveniles are yellow with dark stripes, while adults are browner and have white bands, with the last vestiges of yellow on their caudal fin. This tang is a strong swimmer that requires a large tank with room to comfortably move around.

2 YELLOW TANG
Zebrasoma flavescens
Size: **8 in. (20 cm)**

This tang needs a tank complete with hiding places so that it can mark out its territory.

Origin:	Number of fish per aquarium:	Diet:
Pacific Ocean	1	Algae, small live and frozen foods, dried food that is predominantly plant-based

3 DESJARDIN'S SAILFIN TANG
Zebrasoma desjardinii
Size: **12 in. (30 cm)**

This tang is very aggressive towards members of its own species; it opens out its fins to impress this upon them.

Origin:	Number of fish per aquarium:	Diet:
Pacific Ocean	1	Dried food, small live and frozen foods

3

Triggerfish (Balistidae family)

Triggerfish are able to lock the first spine on their dorsal fins in an upright position to prevent larger fish from eating them. This strategy also makes triggerfish very difficult to catch in a landing net! They feed on invertebrates, so it is out of the question to keep the two together.

Clown triggerfish

It has sturdy teeth for eating small crustaceans and molluscs – the same teeth can also damage the tank decor. It is aggressive toward its own species, but placid with other fish.

1 CLOWN TRIGGERFISH
Balistoides conspicillum
Size: **20 in. (50 cm)**

Most of these triggerfish do not grow more than 12 in. (30 cm) long in an aquarium.

Origin:	Number of fish per aquarium:	Diet:
Indo–Pacific	1	Live and frozen foods, mussels

1

Picasso triggerfish

This is an easy species to keep. It is fairly placid, other than with fish from its own family. It marks out a territory, sometimes disturbing the decor in the process, and takes refuge in a hiding place at night. It can even flatten itself to slide under a rock.

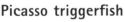

2

2 PICASSO TRIGGERFISH
Rhinecanthus aculeatus
Size: **10 in. (25 cm)**

This species owes its common name to its coloring, which is as bold as a Picasso painting.

Origin:	Number of fish per aquarium:	Diet:
Indo–Pacific	1	Live and frozen prey, mussels

Queen triggerfish

This fish can easily be acclimatized while still young, with the help of fresh or frozen foods. Given its generous size – up to 20 inches (50 cm) – it requires a large tank, ideally with a coral reef rockscape. It will not tolerate members of its own species, but does accept the presence of fish of its own size. It is beneficial to add vegetable food to its basically carnivorous diet.

Red-toothed triggerfish

This triggerfish requires a large tank furnished with hiding places so that it can mark out its territory. Avoid putting it with small fish as it may eat them. It is easy to acclimatize and can live in an aquarium for over 10 years. It has been bred in captivity, but only in extremely large tanks. The male digs a nest for the eggs; these are then aerated and defended by both parents.

3 QUEEN TRIGGERFISH
Balistes vetula

Size: **20 in. (50 cm)**

This is one of the most striking triggers, but it is rarely available commercially.

Origin:	Number of fish per aquarium:	Diet:
Caribbean	1	Meat, mussels, fish, shrimp, small live and frozen foods

4 RED-TOOTHED TRIGGERFISH
Odonus niger

Size: **16 in. (40 cm)**

It is fairly placid with fish of the same size, but not with other triggerfish.

Origin:	Number of fish per aquarium:	Diet:
Red Sea, Indo-Pacific	1	Small live and frozen foods, crustaceans, meat, fish

Boxfish, cowfish and porcupinefish
(Ostraciidae, Diodontidae and Tetraodontidae families)

These are not good swimmers and some are extremely unfriendly toward members of their own family. Some inflate themselves with water to appear larger to their enemies, while others release toxins to defend themselves. They are endearing, however, and may even feed out of your hand – but watch out for their teeth!

Longhorn cowfish

This a very peaceful and socia-ble fish that lives alone and is easy to rear. It must be kept apart from more lively species, however, as it as a slow eater and could miss out on its share of food. The position of its mouth indicates that it finds its food on the substrate. When anxious or disturbed, it releases toxins – a fairly rare phenomenon in fish.

1

1 LONGHORN COWFISH
Lactoria cornuta (Ostraciidae family)
Size: **12 in. (30 cm)**

This fish obviously owes its common name to the two horns on its head.

Origin:	Number of fish per aquarium:	Diet:
Red Sea, Pacific Ocean	1	Small live and frozen foods, molluscs

2 WHITE-SPOTTED BOXFISH
Ostracion meleagris (Ostraciidae family)
Size: **6 in. (15 cm)**

This fish uses only its dorsal and caudal fins to swim.

Origin:	Number of fish per aquarium:	Diet:
Indo-Pacific	1	Small live and frozen foods, molluscs

White-spotted boxfish

This is one of the smallest boxfish. The male's spots stand out against a two-toned background; in the female, the background is uniform. This fish is peaceful and tolerates all other species, so it is easy to acclimatize. It feeds on bottom-dwelling invertebrates.

2

Long-spine porcupinefish

In the event of danger, this fish swells up with water and raises the spines that cover its entire body – it can double its size in this way. If it is inflated with air instead (for example, while being lifted from the water) it will be very difficult for it to deflate itself afterward. Its "beak" (made up of four teeth joined together in two pairs) enables it to crush the shells of crabs, molluscs and even sea urchins. It is closely related to *Diodon hystrix* (the spotfin porcupinefish). Both can become accustomed to eating out of a fishkeeper's hand. Crunching shellfish helps to keep their teeth worn down.

Dog-faced puffer

It is peaceful with fish of its own size, except for other *Arothron*. The white-spotted puffer (*A. hispidus*) and the guineafowl puffer (*A. meleagris*) are among the more common *Arothron* species. They all have the same characteristics and should not be kept with invertebrates.

3 LONG-SPINE PORCUPINEFISH

Diodon holacanthus (Diodontidae family)

Size: **16 in. (40 cm)**

This fish is aggressive with members of its own family but unassertive toward other species.

Origin:	Number of fish per aquarium:	Diet:
Widespread in all warm seas	1	Small live and frozen foods, mussels, shrimp, crabs

4 DOG-FACED PUFFER

Arothron nigropunctatus (Tetraodontidae family)

Size: **12 in. (30 cm)**

Fish of the *Arothron* genus secrete a highly toxic poison: tetrodoxin.

Origin:	Number of fish per aquarium:	Diet:
Red Sea, Indo-Pacific	1	Small live and frozen prey, mussels

Blennies and gobies
(Blenniidae, Gobiidae and Microdesmidae families)

These are small fish that live on the substrate. Blennies have one long dorsal fin, while the gobies have two. In gobies, the two pelvic fins are fused together in the form of a suction disc, which enables them to stick to rocks and avoid being swept away by a current.

Bicolor blenny

Despite its name, this blenny can be completely brown; as well, its coloring alters during the breeding period. It has two small strips of skin on the top of its head. Canine teeth are a valuable defence. It is a poor swimmer and moves about on the bed or among rocks in small leaps.

1 BICOLOR BLENNY
Ecsenius bicolor (Blenniidae family)
Size: **4 in. (10 cm)**

This sociable but timid fish does not enjoy the company of more boisterous species.

Origin:	Number of fish per aquarium:	Diet:
Indo-Pacific	1, or more in a large aquarium	Small live and frozen foods, algae

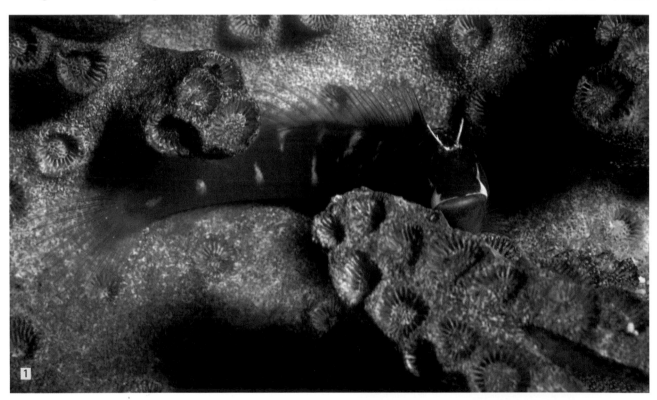

2 BLUE-BANDED GOBY
Valenciennea strigata (Gobiidae family)
Size: **7 in. (17.5 cm)**

The rays of its first dorsal fin are elongated, and it has a horizontal blue band.

Origin:	Number of fish per aquarium:	Diet:
Indo-Pacific	1 or 1 pair	Small live and frozen foods, dried food

3 DECORATED DARTFISH (ELEGANT FIREFISH)
Nemateleotris decora (Microdesmidae family)
Size: **4 in. (10 cm)**

This fish is neither a blenny nor a goby but belongs to a closely related family.

Origin:	Number of fish per aquarium:	Diet:
Indo-Pacific	2 or 3	Small live and frozen foods, dried food

Blue-banded goby

In common with all gobies, this fish has two dorsal fins. These fins differentiate gobies from blennies, which have only one. The blue-banded goby is a calm fish that does not attack other fish or invertebrates. It communicates with fellow gobies by moving its mouth, but it is not known whether this creates an audible signal or is merely a visual display. This substrate-chewing species sifts through particles searching for tiny worms and crustaceans to eat.

Decorated dartfish (elegant firefish)

In this beautifully colored fish, the first rays of the front dorsal fin are permanently erect. It is an easy species to acclimatize with live foods. It digs itself into the sand to hide or takes shelter among the decor; at other times, it likes swimming in slightly turbulent water. A closely related species, the fire goby (*N. magnifica*) boasts an even longer front dorsal fin. Both these fish are somewhat timid and do not appreciate the presence of more lively species.

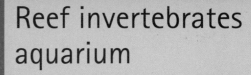

1 DISK (MUSHROOM) ANEMONE
Discosoma spp.

Size: 2–4 in. (5–10 cm)

Origin:	Number of specimens per aquarium:	Lighting:
Indian and Pacific Oceans	1 colony per 65 gal. (250 L)	Moderate to strong

2 GIANT CLAM
Tridacna spp.

Size: 8–40 in. (20–100 cm)

Origin:	Number of specimens per aquarium:	Lighting:
Red Sea, Indian and Pacific Oceans	1 giant clam per 25 to 50 gal. (95 to 190 L)	Intense

3 LEATHER CORAL
Sarcophyton spp.

Size: up to 24 in. (60 cm)

Origin:	Number of specimens per aquarium:	Lighting:
Indian and Pacific Oceans	1 per 50 gal. (190 L)	Intense

4 THORNY BUSH CORAL
Seriatopora hystrix

Size: 8–12 in. (20–30 cm)

Origin:	Number of specimens per aquarium:	Lighting:
Red Sea, Indian and Pacific Oceans	1 foot (12 cm) of coral per 13 gal. (50 L)	Intense

5 HAMMER CORAL
Euphyllia ancora

Size: 12–16 in. (30–40 cm)

Origin:	Number of specimens per aquarium:	Lighting:
Indian and Pacific Oceans	1 colony per 80 gal. (300 L)	Moderate to strong

6 BUBBLE CORAL
Plerogyra sinuosa

Size: around 16 in. (40 cm)

Origin:	Number of specimens per aquarium:	Lighting:
Indian and Pacific Oceans	1 colony per 50 gal. (190 L)	Moderate to strong

2

Miscellaneous species

Some saltwater fish families are only represented by one or two species that may nevertheless be of interest to aquarists.

Banggai cardinalfish

This peaceful, calm fish must be reared with species of the same size and a similar temperament in a tank with shady hiding places, as it prefers being in the dark. Its main asset is the fact that it can be bred with few problems. After a courting display, the eggs are laid and then the male incubates them in his mouth for about three weeks (it appears that the female can also play this role on occasion). When the eggs hatch, the fry remain in the parent's mouth for about another week; after that, they take refuge between the long, thin spines of a diadema sea urchin (particularly *Diadema setosum*). The fry measure approximately $1/2$ inch (1.25 cm), and they are easy to feed with brine shrimp nauplii.

1 BANGGAI CARDINALFISH
Pterapogon kauderni (Apogonidae family)

Size: **3 in. (7.5 cm)**

This species was only discovered in 1990, but it is becoming increasingly common and popular with aquarists.

Origin:	Number of fish per aquarium:	Diet:
Banggai Islands (Indonesia)	2 or 3, or more if there is sufficient space	Small live and frozen foods, mussels, dried food

2 PAJAMA CARDINALFISH
Sphaeramia nematoptera (Apogonidae family)

Size: **3 in. (7.5 cm)**

This peaceful fish must not be kept with larger, livelier species.

Origin:	Number of fish per aquarium:	Diet:
Australia, from Indonesia to Japan	2 or 3	Small live and frozen foods, dried food

2

Pajama cardinalfish

This fish has large eyes, an indication that it is largely nocturnal. It often hides during the day and comes out to feed after the lighting has been turned off. It can be bred in an aquarium, but this is tricky to do. The male keeps the eggs in its mouth until they hatch; the main problem is feeding the fry, which require very small food. Other species from the same family are sometimes available at aquarium stores.

Marine (saltwater) betta

Predators are disconcerted by this fish's fins, which can wrap around its entire body, and by the black patch on its dorsal fin. It swims gently and often remains immobile. It is not advisable to keep it with small fish or shrimp, as it may find them too enticing.

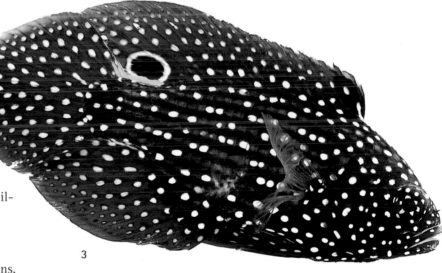

3

3 MARINE (SALTWATER) BETTA
Calloplesiops altivelis (Plesiopidae family)
Size: **7 in. (17.5 cm)**

This fish is somewhat timid and cannot tolerate members of its own species.

Origin:	Number of fish per aquarium:	Diet:
Indian and Pacific Oceans	1	Small live and frozen foods, dried food

Blotched foxface

This fish is peaceful and swims around virtually all the time in search of food. When it is frightened, it hides, its coloring deepens and dark patches appear on its body. Beware! The rays of its dorsal fin are venomous and its sting is painful, although a good deal less dangerous than that of a lionfish. *Siganus unimaculatus* is often confused with a closely related species, *S. vulpinus* (foxface), which behaves in a similar way.

1 BLOTCHED FOXFACE
Siganus unimaculatus (Siganidae family)
Size: **8 in. (20 cm)**

This is a robust fish that can live in a small group. It particularly appreciates plant-based food.

Origin:	Number of fish per aquarium:	Diet:
Indo-Pacific	1 or several, depending on the size of the tank	Small live and frozen foods, algae

1

Longnose hawkfish

This fish barely swims and only moves around in small jumps. Most of the time, it rests on a piece of aquarium decor – sometimes a coral branch – quietly waiting for suitable prey to pass by. For this reason, do not keep it with mobile invertebrates such as prawns or small fish. It is peaceful, except with fish of its own species, as it is inclined to be territorial.

2 LONGNOSE HAWKFISH
Oxycirrhites typus (Cirrhitidae family)
Size: **5 in. (12.5 cm)**

This fish is prized for its red checkerboard coloring.

Origin:	Number of fish per aquarium:	Diet:
Red Sea, Indian Ocean, some parts of the Pacific Ocean	1	Small live and frozen foods, pieces of mussel, dried food

2

Yellow-headed jawfish

Ideally, this fish should be provided with a substrate at least 6 inches (15 cm) deep complete with pebbles, as it digs a burrow by shifting the substrate with its mouth. When it takes refuge, only its head sticks out. It is advisable to mince up its food and scatter this close to the den. It does not attack invertebrates, and prefers to be kept with placid fish of a similar size.

Mandarinfish

This fish lives on the substrate, where it rummages for food. It is a placid fish and easy to rear, although males will not tolerate each other. They can be recognized by the elongation of the rays on the first dorsal fin. There is also a closely related species, *Synchiropus picturatus*, whose gaudy coloring has earned it the nickname of "psychedelic fish." Both species grow to 2.5 inches (6 cm).

3

3 YELLOW-HEADED JAWFISH
Opistognathus aurifrons
(Opistognathidae family)

Size: 5 in. (12.5 cm)

This highly colorful fish lives in a burrow.

Origin:	Number of fish per aquarium:	Diet:
Caribbean	1, or more in a suitably equipped tank	Small live and frozen foods, pieces of mussel

4 MANDARINFISH
Synchiropus splendidus (Callionymidae family)

Size. 2¹/₂ in. (6 cm)

This fish owes its name to its coloring, reminiscent of the clothing of ancient Chinese dignitaries.

Origin:	Number of fish per aquarium:	Diet:
Pacific Ocean	1 or 2	Small live and frozen foods, pieces of mussel

A tropical seahorse

The seahorses of temperate climes are well known, but it is often forgotten that they also live in the tropics. The spotted (or yellow) seahorse *(Hippocampus kuda)* is sometimes available commercially; it can be instantly identified by its bright coloring. It lives in a small group, and often clings with its tail (in fact a caudal fin) to a piece of tank decor or algae. It is peaceful and must be kept with tranquil species. It requires live prey.

4

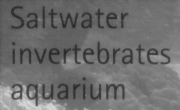

Saltwater invertebrates aquarium

1 CLEANER SHRIMP
Lysmata amboinensis
Size: 2¹/₂ in. (6.5 cm)

Origin:	Number of shrimp per aquarium:	Diet:
Indian Ocean and South Pacific	1 shrimp per 25-gal. (95 L) tank	Omnivorous; scavengers; parasites

2 FIRE (BLOOD) CLEANER SHRIMP
Lysmata debelius
Size: 1¹/₂–2 in. (4–5 cm)

Origin:	Number of shrimp per aquarium:	Diet:
Indian and Pacific Oceans	1 pair per tank	Omnivorous; scavengers

3 PEDERSON'S CLEANER SHRIMP
Periclimenes pedersoni
Size: ¹/₂–1 in. (1.25–2.5 cm)

Origin:	Number of shrimp per aquarium:	Diet:
Southern Caribbean	1 or 2 colonies per 25 gal. (95 L)	Scavengers; parasites

4 PORCELAIN CRAB
Neopetrolisthes maculatus
Size: ¹/₂–1¹/₂ in. (1.25–4 cm)

Origin:	Number of crabs per aquarium:	Diet:
Indian and Pacific Oceans	2 per anemone in 25-gal. (95 L) tank	Plankton

5 PENCIL SEA URCHIN
Heterocentrotus spp.
Size: up to 12 in. (30 cm)

Origin:	Number of urchins per aquarium:	Diet:
Indian Ocean and South Pacific	1 per 80-gal. (300 L) tank	Herbivorous

6 NECKLACE SEA STAR
Fromia monilis
Size: 3–4 in. (7.5–10 cm)

Origin:	Number of crabs per aquarium:	Diet:
Red Sea, Indian Ocean	1 per 80-gal. (300 L) tank	Corals

AQUARIUM PLANTS

Plants are generally considered to play a secondary role in an aquarium, as a support to the fish, although plant-only tanks are also common – often called Dutch aquariums. However, an aquarium is governed by the laws of nature, so a subtle balance must be achieved between the water, light, plants and fish. Apart from their decorative aspect – undoubtedly the foremost concern of many aquarists – plants play an essential role in the stability of a tank, as they give off oxygen and consume substances that are harmful to fish. In fact, plants play several roles – apart from their esthetic and biological functions, they also offer fish a place to shelter. Dense areas of vegetation, particularly at the rear of the tank, enable them to take refuge and feel safe, as well as provide an ideal location for spawning. Plants are also favorite spots for fry; they find tiny morsels of food in them and, above all, convenient places to hide.

The numerous species of plants now available in aquarium stores make it possible to establish thematic biotope aquarium displays that recreate habitats as varied as steamy jungle pools and fast-flowing mountain streams. With care and creative skill, any aquarium hobbyist can now produce underwater displays that are not only visually stunning, but also highly appropriate environments for the fish and other living creatures they support.

The biology of aquarium plants

It is unusual for fish to live without plants in their natural environment, as they provide them with shelter – particularly when they are young – and sometimes also serve as food. It is often forgotten, however, that plants consume carbon dioxide and produce vital supplies of oxygen under the influence of sunlight.

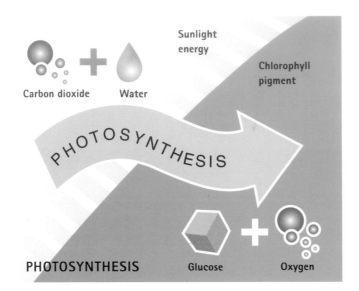

PHOTOSYNTHESIS

Carbon dioxide + Water — Sunlight energy — Chlorophyll pigment — PHOTOSYNTHESIS — Glucose + Oxygen

Photosynthesis

Plants cannot grow without light. Their cells contain structures called chloroplasts, which contain photosynthetic pigments, such as chlorophyll. This traps the energy from light to convert water and carbon dioxide (CO_2) into glucose and oxygen (O_2): this process is known as photosynthesis. Oxygen dissolves in water, but glucose, a simple sugar, remains in the plant and provides it with energy to grow and reproduce; it is also stored in the form of starch (like that in a potato).

Photosynthesis is regulated by a number of environmental factors, particularly light, CO_2 and temperature. These can all be limiting factors to the optimal photosynthetic rate: if one of these three factors does not reach a sufficient level, any increase in the others will not boost photosynthesis activity.

An oxygen deficit in well-planted tanks

Photosynthesis ceases at night, and plants therefore stop producing oxygen. They continue to respire, however, as do the other organisms (fish, invertebrates and also bacteria). Respiration uses oxygen and produces carbon dioxide as a waste product, so the oxygen levels in a heavily planted tank can drop sharply overnight. The only remedy is to aerate the aquarium at night, but only if the fish seem to need it.

Limiting factors for photosynthesis

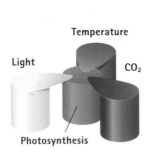

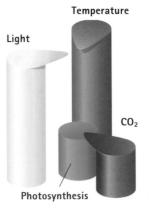

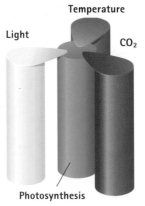

Light, temperature and carbon dioxide (CO_2) are limiting factors of photosynthesis.

Without additional CO_2, increases in light and temperature have no effect on photosynthesis rate.

With added CO_2 in the aquarium, photosynthesis will increase toward a more optimal rate.

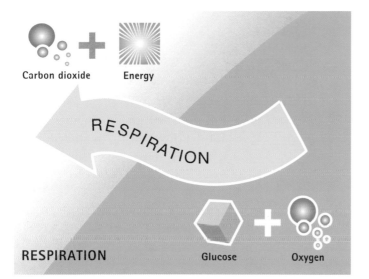

Carbon dioxide   Energy

RESPIRATION

RESPIRATION   Glucose   Oxygen

Plant pigments

Photosynthesis only takes place in the green parts of plants (leaves and stems) – the roots are not involved because light does not reach them. But the green pigment chlorophyll is not the only one that can harness light energy to power photosynthesis. Other plant pigments include the carotenoids, which create reddish colors. These are less effective at photosynthesis, so predominantly red plants must receive more light than green ones. In dull light, the leaves remain green.

Like all living organisms, plants respire continuously, 24 hours a day. The function of respiration is to break down food and release energy into the cells. Oxygen is consumed and carbon dioxide is released as a waste product   a cycle opposite to that of photosynthesis. In an average 24-hour period, the plants produce more oxygen through photosynthesis than they use up in respiration.

This closeup of an aquatic plant shows the ribs and veins that support the structure; they transport water, nutrients and food products around the plant.

THE DAY-NIGHT CYCLE IN A PLANTED AQUARIUM

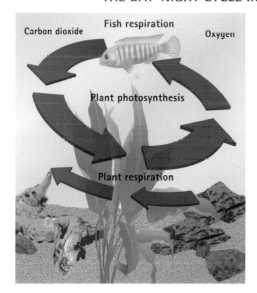

Carbon dioxide   Fish respiration   Oxygen

Plant photosynthesis

Plant respiration

During the day, or under artificial light, both plants and fish respire, using up oxygen and releasing carbon dioxide. Photosynthesis uses carbon dioxide and gives off relatively more oxygen, so the level of oxygen goes up.

At night, photosynthesis stops while respiration in fish and plants continues. Oxygen levels fall and carbon dioxide levels rise. In a well-planted aquarium, surface movement at night will vent off CO_2 and encourage oxygen to dissolve.

Adaptations to an aquatic life

Aquatic plants are perfectly designed for underwater life. The stems and leaves have far less supporting tissue than terrestrial plants, because they rely on the surrounding water for support. A system of air cavities maintains the plant's buoyancy and this, combined with less supportive tissue, allows the plants to be far more flexible underwater, easily bending and moving with the current.

The leaves of aquatic plants are thinner than those of land plants. Thinner leaves are more efficient at photosynthesis, because light is able to penetrate easily to all the cells. Even so, most of the chlorophyll-containing structures (chloroplasts) are found near the upper surface where the light is stronger. This explains why the upper side of a leaf is often greener than the underside.

Unlike terrestrial vegetation, aquatic plants can absorb mineral salts and nutrients through their leaves as well as their roots.

Aquatic plants and water quality

Most aquatic plants can adapt to a wide range of conditions. Some are more delicate, however, and require a specific water quality, favoring soft, acidic water. They die in hard water because of low CO_2 levels.

To maximize light penetration, and therefore photosynthesis, the water must be clear. Effective mechanical filtration will help to keep fine leaves free of debris.

Right:
Cabomba (*Cabomba* spp.) is a true aquatic plant, with thin leaves and stems designed for buoyancy and flexibility under water.

INSIDE A LEAF AND PLANT CELL

These drawings show an enlarged cross-section of a val (*Vallisneria* spp.) leaf, a typical aquatic plant with thin leaves, plus a detailed view inside one plant cell.

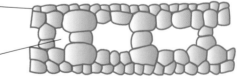

Both surfaces are made up of a layer of cells covered with a thin cuticle.

Air spaces provide support and structure.

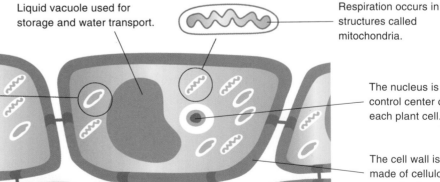

Photosynthesis takes place within the nutrient-rich liquid inside chloroplasts.

Inside each chloroplast, the green pigment chlorophyll is contained in plates that move toward light like solar panels.

Liquid vacuole used for storage and water transport.

Respiration occurs in structures called mitochondria.

The nucleus is the control center of each plant cell.

The cell wall is made of cellulose, a matrix of carbohydrates.

Right:
A dip-strip test such as
this is a quick and easy
way to check on water
quality levels such as
pH and hardness.

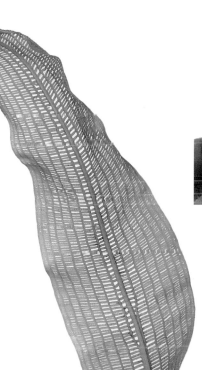

Above: At the right temperature, with high CO_2
levels, strong lighting and a good nutrient supply,
photosynthesis in aquatic plants will work at peak
efficiency. The bubbles of oxygen on these leaves
show that photosynthesis is at a high rate.

Left:
Certain aquatic plants,
such as this Madagascar
laceleaf *(Aponogeton
madagascariensis)*, have
a fine leaf structure. Good
filtration will prevent such
leaves becoming clogged.

Supplying carbon dioxide

One of the most common causes of
unsuccessful plant-keeping is a lack of carbon
dioxide. There are several ways of boosting the
levels of this gas in a planted aquarium. For
small tanks, there is a system based on mixing
yeast with a sugar solution and piping the carbon
dioxide gas produced to a small dosing device
in the tank. For larger tanks, the best way is to
install a system that releases controlled amounts
of gas from a compressed cylinder. Since extra
CO_2 is only needed during the day, these systems
operate only when tank lighting is on.

The substrate

The substrate has many functions in an aquarium. It provides a suitable medium
for holding plants and decor in place and creates an attractive display, but
for plants and their development, a good substrate will provide much more.

A planted aquarium should have a mixture of substrates, all of which play a part in providing the best overall conditions for growth.

Gravel and sand

Standard aquarium gravel, also called pea gravel due to its rounded, smooth appearance, is available in a number of grades (particle size) from $\frac{1}{12}$–$\frac{1}{2}$ inch (2–10 mm). The smaller grades of pea gravel will provide good support and are reasonably efficient rooting mediums, but they offer little nutritional value to foster good plant growth.

Small-grade, about $\frac{1}{12}$ inch (2 mm), lime-free gravel is a better starting point as a main rooting substrate, because it provides excellent root support and is totally inert, without the lime and calcareous contaminants in standard gravel that may affect water quality.

As an alternative to gravel, sand can look very attractive in an aquarium, but problems arise if it used as the sole substrate. Because of its small particle size, it can compact over time, leading to bad water circulation and oxygen-deprived areas. These stagnant areas can release small quantities of noxious gases that may harm the livestock in the aquarium. Used as a layer or with heating cables, however, sand can be a good choice.

Clay-based and enriched substrates

Clay-based substrates, usually called laterites, are rich in iron and release nutrients over a long period of time. Other types of enriched substrates are also available from aquarium stores. The best way to use all these substrates is to mix them with lime-free gravel or lay them down as a thin layer between two other substrates. This not only makes the nutrients they contain available to plant roots, but also prevents them from floating free in the aquarium and muddying the water.

Preparing and adding the substrate

All types of gravel must be rinsed with clean water before use to remove any dust and debris. Even with gravel pre-washing, you may find that the aquarium will cloud slightly when you add water. This will clear in 24 hours. If a single type of substrate is being used, spread it on the bottom of the aquarium, with a slight backward slope – 2–3 inches (5–7.5 cm) thick to the rear, 1–2 inches (2.5–5 cm) at the front. If you are using a nutrient-rich material, mix it with or sandwich it between other substrates as mentioned above. If you are using a heating cable, place this on the tank base and build up substrate layers to the right depth.

Right:
Laying an enriched substrate to boost plant growth. Cover the base of the tank with fine sand 2 inches (5 cm) deep, topped with a thin layer of iron-rich laterite substrate. Then add a top layer of lime-free quartz gravel up to 4 inches (10 cm) deep.

Aquarium lighting

With a few rare exceptions, aquarium plants need fairly strong light to flourish. Aquarium lighting must provide light of the right color, strength and duration to simulate the natural sunlight available to them in their original tropical habitats.

Fluorescent bulbs are the most widely used source of light in aquariums with a water depth of less than 20 inches (50 cm). They are very efficient, use little electricity and are relatively cheap. Also, they do not heat the water and can last for six to twelve months with little reduction in quality. They are available in various lengths, which correspond to their power ratings. (The box on this page provides guidance on how many and what type of bulbs to use.)

Fluorescent bulbs designed to boost aquarium plant growth were originally developed for horticultural use and produce a pink glow, which gives aquariums a slightly garish look. To remedy this, full spectrum daylight bulbs are added to balance the light.

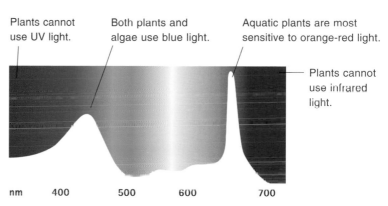

Plants cannot use UV light.

Both plants and algae use blue light.

Aquatic plants are most sensitive to orange-red light.

Plants cannot use infrared light.

nm   400   500   600   700

nm: light wavelengths in nanometers (billionths of a meter).

The lighting period is a vital factor in promoting good plant growth. Many aquarium plants are from tropical regions and need good light for 12 to 14 hours a day. A timer is a very practical accessory – ideally, each bulb or lamp should have its own timer switch, so that the lighting can be built up and faded gradually. This avoids a sudden day/night change, which may upset fish.

Left:
Suspending a metal-halide or mercury vapor lamp above the aquarium is good way of providing the intense light that some plants need.

For deeper aquariums, metal-halide or mercury vapor lamps should be suspended over the tank, at least 12 inches (30 cm) above. These are strong enough to allow light to penetrate right to the bottom of the aquarium.

NUMBER AND TYPE OF FLUORESCENT BULBS

The number of bulbs must be proportional to the dimensions of the aquarium. After the first two, different colors can be added.

Tank dimensions (length x width x depth)	Tank volume	Bulb length/ power	Total number of bulbs	Horticultural bulbs/ daylight bulbs
24 x 12 x 12 in. (60 x 30 x 30 cm)	14 gal. (54 L)	18 in. (45 cm) /15 W	2	1/1
32 x 12–14 x 12 in. (80 x 30–35 x 30 cm)	25–30 gal. (95–110 L)	24 in. (60 cm) /20 W	2	1/1
40 x 16 x 16 in. (100 x 40 x 40 cm)	45 gal. (170 L)	36 in. (90 cm) /30 W	3	1/2
48 x 16–18 x 20 in. (120 x 40–45 x 50 cm)	63 gal. (240 L)	36 in. (90 cm) /30 W	4	1/3 or 2/2
60 x 20 x 20 in. (150 x 50 x 50 cm)	100 gal. (375 L)	48 in. (120 cm) /40 W	4 or 5	2/2 or 2/3

263

Choosing and using plants

There are several aspects to consider when choosing plants for your aquarium. First of all, you must choose plants that will grow in your water conditions: most will thrive in soft, slightly acidic water, although some demand harder and more alkaline conditions. Then, you must consider the amount of light they need and provide the right level of illumination to boost their growth. Finally, you should choose plants suited to your experience, matching their hardiness with your ability to grow them well. Above all, be sure to choose the size, shape, texture and color of plants that will make your display the envy of all your friends.

Right:
This Java fern (*Microsorium* spp.) has been attached to a piece of bogwood, where it will develop roots and grow.

To create an attractive planted display, choose fairly big plants for the background at the rear and sides of the tank, medium-sized ones for the midground areas and small or low-growing ones for the foreground. Group plants of the same variety together for best effect. It is not necessary to have a large number of species – four or five are sufficient in a 25-gallon (95 L) tank.

Above and right:
Before planting, remove the lead weight and separate the individual plants.

Buying the plants

Good-quality plants are easy to recognize. They should have strong vibrant colors – except for some crypts (*Cryptocoryne* spp.) that are naturally brown and drab – and show signs of new growth. Avoid buying plants with many leaves that are damaged, yellowing, excessively brown or black. Some plants are sold in bunches of cuttings or rooted divisions held together with a lead strip. Remove this before planting. Others are sold in perforated plastic pots containing rockwool or a similar support material. Tease this away from the roots before planting.

Plants with plenty of root growth may need trimming. In fact, it is a good idea to cut roots back to avoid damaging them during the planting process – they will quickly regenerate. It is also important to check new plants for snails, and also look closely at the leaves and stems for gelatinous blobs containing snail eggs.

Above:
Some plants are supplied as cuttings embedded in rockwool. Gently remove the bunch from its plastic pot.

Right:
Remove the rockwool carefully from plants with developed roots.

Acclimatizing plants

Once they have been put in place, plants will start to straighten up in a few days and grow toward the light source, although it may take a couple of weeks for the roots to develop properly. It is only after this phase that the plants will start to grow – but do not be surprised if the form and color of the new leaves are slightly different from the originals. The plant will adjust to its new conditions: water quality, nutrition and lighting levels.

Planting techniques

The best method for planting is to make a small hole with your fingers, put the base of the plant into it and replace the substrate around it. If the substrate contains an enriched layer, make sure that none of it rises to the surface. When planting cuttings without roots, strip off the lower leaves before burying the stem in the substrate. Insert rooted plants into the substrate up to the base of the first leaves, which may sometimes be lacking in pigmentation.

Many plants that naturally grow attached to rocks or wood develop from a rhizome that clings to the surface. Tie this rhizome to the rock or wood with black cotton or nylon thread. New healthy roots will spread out across the support within a few weeks.

Above:
Plants will start to grow one to two weeks after their introduction into the tank.

Choosing the best plant

The plant profile section (starting on page 270) features plants for the background, midground and foreground areas of the aquarium, plus a brief survey of floating and specimen plants.

The following information is provided:

- The common name
- The scientific name (possibly with a synonym)
- The geographical origin

- The maximum size that it can achieve in ideal conditions
- The intensity of light required
- The optimum temperature
- The difficulty of cultivation:
1 Robust, resistant plant, easy to grow – recommended for beginners
2 Fairly easy to keep, but may need extra care and conditions to flourish
3 More delicate plant that may have specific needs to achieve good growth

Above:
The best planting technique is to make a small hole in the substrate with two fingers, gently push the plant into it and then replace the substrate around it. To create a more attractive display, place the same plants close together.

Feeding and propagating plants

In order to grow properly and remain in good health, aquarium plants need carbon dioxide to "fuel" photosynthesis and a number of minerals and organic nutrients, which they take in either through their leaves or roots. Supplying the correct fertilizers can be thought of as giving plants a good, balanced diet. Healthy plants will multiply by various means.

If you have laid down a layer of nutrient-rich material in the substrate (see page 262), there is no need for added fertilizer. Remember, however, that this enrichment will eventually be exhausted. Most slow-release, nutrient-rich substrates will lose part or all of their nutrients within three years, at which point they will need replacing.

Most aquarists use a standard gravel substrate, which has no nutritional value, so it is necessary to provide extra food by adding plant fertilizers. The easiest option is to use a liquid plant food that you add regularly to the water, following the recommended dosage. The best time to add liquid fertilizer is just after a partial water change, thus ensuring good nutrient levels. Tablet fertilizers are also available. These are particularly convenient for feeding one or a few plants. Simply bury them at the base of the plants, near the roots.

Above:
This liquid plant food has a measuring cup incorporated into the lid for accurate dosing. Always follow the directions provided.

Right:
Slowly add diluted liquid fertilizers to the tank. Be careful not to overdose, which could cause algal blooms.

Place fertilizer tablets close to the roots.

Propagating aquarium plants

Although many aquatic plants will produce flowers above the surface, pollination and seed production is a difficult and unreliable way of propagating them. Fortunately, there are many techniques of asexual, or vegetative, propagation that work well.

Plants with runners

Plants such as Vals (*Vallisneria* spp.), Amazon swords (*Echinodorus* spp.), and crypts (*Cryptocoryne* spp.) produce horizontal stemlike shoots called runners that carry daughter plants. In the wild, these small plants root into the substrate and the runner breaks down. In the aquarium, simply allow these small plants to develop a good growth of leaves and roots and then cut the runner that joins them to the parent plant.

Taking cuttings

Just like garden plants, aquatic plants with a central stem can be propagated by taking cuttings. These can be taken from the side

Take top cuttings by snipping between nodes.

Remove the lower leaves.

Insert to the level of the lower leaves.

Divide rhizomes into pieces, each with a shoot.

Plantlets can develop at the ends of leaves. Detach and plant separately.

shoot or from the top and middle sections of the plant. Top cuttings will establish and grow more quickly. Using a pair of sharp scissors, cut off the top section of a stem and remove any leaves from the bottom one or two nodes (the points at which the leaves arise). Insert the cutting into the substrate, covering these bottom nodes. Roots and new side shoots should develop in a week or two.

Dividing rhizomes

Some plants grow from a modified stem called a rhizome. This looks like a thick root at the base of the plant. To produce new plants, carefully lift out the original plant and cut the rhizome into sections with a sharp knife, ensuring that each division has at least one good shoot. For those plants, such as Java fern (*Microsorium* spp.), that grow attached to rocks or wood, leave the original plant in place as you remove sections of rhizome. Attach these to new supports.

Propagating from offsets

Quite a few aquarium plants grow in clumps. These form as new plants from around the base of the original parent. Simply separate these and plant elsewhere.

Below:
Feeding plants well and keeping their growth under control can produce stunning displays like this.

A Dutch aquarium

1 ONION PLANT
Crinum thaianum

Size: up to 5 ft. (1.5 m)

Origin:	Number of plants per aquarium:	Lighting:
Thailand	1 per 40-gal. (150 L) tank	Bright

2 URUGUAY AMAZON SWORDPLANT
Echinodorus uruguayensis

Size: 12 in. (30 cm) or more

Origin:	Number of plants per aquarium:	Lighting:
Western South America	1 to 3 per 40-gal. (150 L) tank	Bright

3 PENNYWORT
Hydrocotyle verticillata

Size: 10 in. (25 cm)

Origin:	Number of plants per aquarium:	Lighting:
North and Central America	1 per 20-gal. (75 L) tank	Bright

4 TIGER LOTUS
Nymphaea lotus

Size: over 20 in. (50 cm)

Origin:	Number of plants per aquarium:	Lighting:
East Africa and Asia	1 per 40-gal. (150 L) tank	Bright

5 GLOSSOSTIGMA ELATINOIDES
Glossostigma elatinoides

Size: under 1 in. (2.5 cm)

Origin:	Number of plants per aquarium:	Lighting:
Australia, New Zealand	5 pots per 25 gal. (95 L)	Very bright

6 HEMIANTHUS
Hemianthus micranthemoides

Size: 6–8 in. (15–20 cm)

Origin:	Number of plants per aquarium:	Lighting:
Southern United States, Cuba	3 to 5 per 20 gal. (95 L)	Moderate to very bright

4

6

Background plants

These are big, and some of them grow very quickly. Species that readily put out suckers can form attractive clumps; others stand out better if only a single plant is used. All these plants are ideal for the rear and sides of the aquarium.

1

1 RUFFLED AMAZON SWORDPLANT
Echinodorus major

Maximum size: **20–24 in. (50–60 cm)**

Origin:	Light:	Temperature:	Difficulty:
Brazil	Bright	72–82°F (22–28°C)	1

2 BROADLEAF AMAZON SWORDPLANT
Echinodorus bleheri (E. paniculatus; E. rangeri)

Maximum size: **20–24 in. (50–60 cm)**

Origin:	Light:	Temperature:	Difficulty:
South America	Moderate to bright	72–82°F (22–28°C)	1

Crypt

It tolerates hard water and should be planted in a group to the rear or at the sides of the tank. In good conditions, shoots will grow from the base; these can then be cut off and planted elsewhere in the aquarium.

3 CRYPT
Cryptocoryne balansae

Maximum size: **14–16 in. (35–40 cm)**

Origin:	Light:	Temperature:	Difficulty:
Vietnam, Thailand	Moderate to bright	73–82°F (23–28°C)	2 or 3

Ruffled Amazon swordplant

As its species name suggests, this can grow very big and so is more appropriate for a large tank. Although most of the species from this genus are amphibious bog plants, the giant *Echinodorus* species is totally aquatic. It is easy to cultivate, preferably as a single specimen. It needs plenty of space, as one stem can produce up to 20 leaves. It prefers soft, slightly acidic water, but can tolerate a degree of hardness. Its growth can be enhanced by the use of fertilizers. It reproduces by means of daughter plants on runners.

Broadleaf Amazon swordplant

This is ideal in large tanks, either in a well-spaced group or on its own, as is a related species, *Echinodorus amazonicus*. It grows quite slowly, but in good conditions it can be eye-catching in an aquarium.

2

3

Green cabomba

This must be grown in water entirely free of suspended matter, as this could overwhelm its delicate foliage. Plant in groups of at least five to create a decorative effect. Cuttings grow very easily. There are several closely related species, including the red cabomba (*Cabomba piauhyensis* or *C. furcata*) and the yellow, or giant, cabomba (*C. aquatica*).

Giant ambulia

This needs soft, slightly acidic water, along with a rich substrate or fertilizer. It will be less bushy in hard water and under weak lighting. If the top is pruned, it will grow side shoots that can be cut off and replanted.

5 GIANT AMBULIA
Limnophila aquatica
Maximum size: **20 in. (50 cm)**

Origin:	Light:	Temperature:	Difficulty:
Southeastern Asia	Bright	72–80°F (22–27°C)	3

4 GREEN CABOMBA
Cabomba caroliniana
Maximum size: **up to 40 in. (100 cm)**

Origin:	Light:	Temperature:	Difficulty:
Central America, northern South America	Bright	61–79°F (16–26°C)	1 or 2

Green myriophyllum (water milfoil)

Use several slightly spaced out sprigs to form a clump. The water must be well filtered to avoid any detritus settling on the fine foliage. There are several closely related species of green-leafed milfoils, which live in neutral or slightly alkaline water. Their soft leaves are much appreciated by herbivorous fish.

Red milfoil

In bright light, this plant's red coloring contrasts strongly with other vegetation. Space out the stems to allow light to reach the lower leaves.

6 GREEN MYRIOPHYLLUM (WATER MILFOIL)
Myriophyllum hippuroides
Maximum size: **20 in. (50 cm)**

Origin:	Light:	Temperature:	Difficulty:
North and Central America	Bright	59–73°F (15–23°C)	2

7 RED MILFOIL
Myriophyllum mattogrossense (M. tuberculatum)
Maximum size: **24 in. (60 cm)**

Origin:	Light:	Temperature:	Difficulty:
South America	Very bright	72–83°F (22–28°C)	3

Red hygrophila (copperleaf)
The lower surface of this plant's leaves is darker than the upper one. It prefers water that is slightly acidic and very soft. For optimum effect, it must be planted in a group, with space between each stem. It can be propagated from cuttings.

Dwarf hygrophila
This plant is often found in aquariums as it can adapt to various types of water. It must be planted in a group of at least four stems. Take cuttings regularly.

1 RED HYGROPHILA (COPPERLEAF)
Alternanthera reineckii (A. rosaefolia)
Maximum size: **20 in. (50 cm)**

Origin:	Light:	Temperature:	Difficulty:
South America	Very bright	72–79°F (22–26°C)	3

2 DWARF HYGROPHILA
Hygrophila polysperma
Maximum size: **20 in. (50 cm)**

Origin:	Light	Temperature:	Difficulty:
India	Bright	64–86°F (18–30°C)	1

Giant red rotala
This plant prefers very soft, acidic water and cannot tolerate strong water flow. Its red coloring will be vivid under good conditions, especially with the addition of iron-rich fertilizer and very strong lighting. Adventitious roots grow spontaneously from the stems, making it easy to take cuttings.

Dwarf rotala
This is easier to tend than its cousin and grows quickly. It must be pruned regularly to enable light to penetrate to the lowest leaves.

3 GIANT RED ROTALA
Rotala macrandra
Maximum size: **16 in. (40 cm)**

Origin:	Light:	Temperature:	Difficulty:
India	Very bright	75–84°F (24–29°C)	3

4 DWARF ROTALA
Rotala indica (R. rotundifolia)
Maximum size: **24 in. (60 cm)**

Origin:	Light:	Temperature:	Difficulty:
Southeastern Asia	Bright to very bright	68–84°F (20–29°C)	1–2

Elodea (pondweed)

This is one of the best known of all aquarium plants, being adaptable and easy to keep. It can adapt to a wide range of conditions, including cold water, but it thrives best in hard water. Propagation using cuttings is a simple affair: side shoots with roots often grow spontaneously – just cut these off carefully and replant them.

5 ELODEA (PONDWEED)
Egeria densa (Elodea densa)

Maximum size: **20 in. (50 cm) or more**

Origin:	Light:	Temperature:	Difficulty:
Worldwide	Bright	Up to 79°F (26°C)	1

Water wisteria

This grows profusely in its native habitat. Its delicately indented leaves are highly decorative. It prefers a nutrient-rich substrate. Propagate it by means of cuttings.

6 WATER WISTERIA
Hygrophila difformis (Synnema triflorum)

Maximum size: **20 in. (50 cm)**

Origin:	Light:	Temperature:	Difficulty:
India, Thailand	Bright	73–82°F (23–28°C)	1 or 2

Straight val

This very popular plant quickly produces young shoots by means of runners. Flowers emerge from a small, spiraled stem, hence the species name. The leaves are straight, unlike a closely related species, *Vallisneria asiatica*, which does not like overly soft water and is available in several varieties.

7 STRAIGHT VAL
Vallisneria spiralis

Maximum size: **16–20 in. (40–50 cm)**

Origin:	Light:	Temperature:	Difficulty:
Widespread in tropical and subtropical regions	Bright	59–86°F (15–30°C)	1

273

Midground plants

These smaller plants are perfectly suited to the sides and center of a tank. They can even be placed against the back glass of shallow aquariums. Some of them benefit from being planted singly as specimen plants, as this sets them apart from their neighbors by highlighting differences in color or the shape of their leaves.

1 WATER HEDGE
Didiplis diandra

Maximum size: 13 in. (33 cm)

Origin:	Light:	Temperature:	Difficulty:
North America	Very bright	68–79°F (20–26°C)	2 or 3

Water hedge

A group of 10 plants will create a visual impact. The small leaves contrast with the larger ones of other plants. It prefers rich soil and strong lighting. The plants reproduce easily: just cut off the side shoots and replant them.

2

Green ludwigia

This is one of several species of *Ludwigia*. It is sold in a number of forms, including a variety with brown-red leaves. They do not demand any exacting conditions to grow and can be propagated easily by taking cuttings.

Creeping Jenny

This is perfect for a coldwater aquarium, as it will not tolerate temperatures above 72°F (22°C). It is not fussy, except with regard to lighting, which must be very bright.

1

3

2 GREEN LUDWIGIA
Ludwigia palustris

Maximum size: 16 in. (40 cm)

Origin:	Light:	Temperature:	Difficulty:
America	Bright	64–79°F (18–26°C)	1–2

3 CREEPING JENNY
Lysimachia nummularia

Maximum size: 6–10 in. (15–25 cm)

Origin:	Light:	Temperature:	Difficulty:
Europe, North America	Very bright	50–72°F (10–22°C)	1

4 BROADLEAF ANUBIAS
Anubias barteri

Maximum size: **13 in. (33 cm)**

Origin:	Light:	Temperature:	Difficulty:
West Africa	Moderate to bright	72–82°F (22–28°C)	2

Broadleaf anubias
This name is freely applied to several closely related species. This is a slow-growing plant with leaves that are too tough to be eaten by herbivorous fish. The top leaves can grow out of the water. One variety can grow on a support (see Specimen plants, page 280).

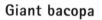

Giant bacopa
This grows quickly and adapts to hard water. If the lighting is very strong, the leaves take on a reddish bronze color. Cuttings can be taken easily, which makes it possible to create an attractive group in a short time. There are several related species that resemble it; all prefer very soft, neutral or slightly acidic water.

Stargrass
When this plant reaches the surface, it produces aerial leaves that are smaller than the submerged ones; in this case, the stem must be cut and replanted. It is sturdy and can form bushy clumps if it is pruned regularly.

5 GIANT BACOPA
Bacopa caroliniana (B. amplexicaulis)

Maximum size: **12 in. (30 cm)**

Origin:	Light:	Temperature:	Difficulty:
Southern and Central U.S.	Bright	68–82°F (20–28°C)	2

6 STARGRASS
Heteranthera zosterifolia

Maximum size: **20 in. (50 cm)**

Origin:	Light:	Temperature:	Difficulty:
South America	Bright	72–79°F (22–26°C)	2

275

Foreground plants

These are ideal for planting at the front of an aquarium, right up against the glass. They can be separated from the larger plants by bogwood or rocks. An impression of depth can be achieved by placing the tallest specimens at the back, and gradually decreasing the size of plants toward the front of the aquarium.

1 BECKETT'S CRYPTOCORYNE
Cryptocoryne beckettii
Maximum size: **6–8 in. (15–20 cm)**

Origin:	Light:	Temperature:	Difficulty:
Sri Lanka	Undemanding	75–82°F (24–28°C)	1 or 2

Undulate cryptocoryne

The edges of this plant's leaves are more wavy than that of other species in this genus. It is easy to keep and very quickly sends out new shoots. It turns a brownish red color under strong lighting.

Beckett's cryptocoryne

This is one of the most common foreground plants. There are several varieties, but they all grow slowly. This cryptocoryne tolerates neutral, fairly hard water. A group of these plants will create a delightful foreground that has the added advantage of requiring little maintenance. It multiplies by means of runners.

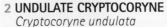

2 UNDULATE CRYPTOCORYNE
Cryptocoryne undulata
Maximum size: **4–8 in. (10–20 cm)**

Origin:	Light:	Temperature:	Difficulty:
Sri Lanka	Undemanding	72–82°F (24–28°C)	1

Cryptocoryne wendtii

This is one of the most popular cryptocorynes; there are several varieties, which differ with respect to their shape and leaf size. It is sturdy and adapts well to a tank if the water quality is stable – it prefers moderate hardness and neutral (or slightly alkaline) conditions. It grows slowly.

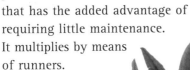

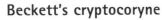

3 CRYPTOCORYNE WENDTII
Cryptocoryne wendtii
Maximum size: **4–6 in. (10–15 cm)** – there is also a larger variety

Origin:	Light:	Temperature:	Difficulty:
Sri Lanka	Moderate	72–82°F (24–28°C)	1

Pygmy chain swordplant

This plant can spread prolifically by means of runners to form a "lawn," which must be thinned out from time to time. It requires neutral or slightly acidic water of negligible hardness. It may sometimes be overrun by filamentous algae.

Hairgrass

This is suited to coldwater tanks. As it is tufted, it traps particles suspended in the water, so a good filtration system is imperative. A closely related species (Eleocharis parvula) is also available.

4 PYGMY CHAIN SWORDPLANT
Echinodorus tenellus
Maximum size: 5 in. (12.5 cm)

Origin:	Light:	Temperature:	Difficulty:
Southern United States and Brazil	Bright	62–80°F (17–27°C)	2

5 HAIRGRASS
Eleocharis acicularis
Maximum size: 4–6 in. (10–15 cm), sometimes more

Origin:	Light:	Temperature:	Difficulty:
Worldwide (tropics)	Bright to very bright	50–77°F (10–25°C)	2–3

Giant sagittaria

With its sturdy leaves, this plant could also be used singly in open spaces, as a midground plant or grouped toward the center. It will grow in a coldwater tank.

Lilaeopsis

It grows quickly to create a dense carpet; it is not fussy about water quality. It is often confused with *Echinodorus tenellus*.

6 LILAEOPSIS
Lilaeopsis novae-zelandiae
Maximum size: 5 in. (12.5 cm)

Origin:	Light:	Temperature:	Difficulty:
New Zealand	Very bright	64–82°F (18–28°C)	2

7 GIANT SAGITTARIA
Sagittaria platyphylla
Maximum size: 8 in. (20 cm)

Origin:	Light:	Temperature:	Difficulty:
Southern United States	Very bright	61–79°F (16–26°C)	2

Floating plants

These plants have the advantage of supplying excellent shelters for small fish and fry. Avoid using them in areas with strong water flow, as there is a risk that the plants will be pushed into one corner of the aquarium.

Hornwort

It grows very quickly and spreads across the surface. It serves as a shelter for fry and can be used to filter the light in specific areas of the tank. It can also be grown planted in the bed or fixed onto a support (such as wood or rock) with nylon wire. A cutting produces a new plant in no time, but it must be handled with care, as it is fragile. The hornwort is suited to both coldwater and tropical aquariums.

Indian fern

This can be planted in the substrate, but the roots should not be buried too deep, otherwise they will rot. It grows better on the surface, where its leaves are not so fine. It is easy to propagate, as small shoots form on the leaves. It is very hardy and a good choice for novice aquarists.

2 INDIAN FERN
Ceratopteris thalictroides
Maximum size: **14 in. (35 cm)**

Origin:	Light:	Temperature:	Difficulty:
America, Africa, Asia, northern Australia	Bright	72–79°F (22–26°C)	1

1 HORNWORT
Ceratophyllum demersum
Maximum size: **16 in. (40 cm)**

Origin:	Light:	Temperature:	Difficulty:
Worldwide	Bright	50–82°F (10–28°C)	2

3 WATER LETTUCE
Pistia stratiotes

Maximum size: **2–4 in. (5–10 cm)**

Origin:	Light:	Temperature:	Difficulty:
Widespread in tropical regions	Bright	73–82°F (23–28°C)	1 or 2

Water lettuce

Fry greatly appreciate the shelter provided by its roots, which can grow to over 4 inches (10 cm). It reproduces rapidly by means of runners that give rise to young plants at their tips. Provide space and ventilation between the water surface and the lights to prevent the leaves from being burned.

Crystalwort

This is a moss that has neither leaves nor roots. It grows so quickly that it can overrun the entire surface of the water. It is a great asset for fry, as it is colonized by infusorians, which often serve as their first food, and it produces plenty of oxygen. It can grow completely submerged; in this case, it must be fixed to a support with nylon wire.

Salvinia auriculata

Leaves $1/2$–1 inch (1.25–2.5 cm) grow from both sides of a horizontal stem; their upper surface has a hairy texture. This plant is easy to cultivate and grows so quickly that it can even become invasive.

4 CRYSTALWORT
Riccia fluitans

Maximum size: **6–12 in. (15–30 cm)**

Origin:	Light:	Temperature:	Difficulty:
Worldwide	Moderate	59–86°F (15–30°C)	1

5 SALVINIA AURICULATA
Salvinia auriculata

Maximum size: **8–10 in. (20–25 cm)**

Origin:	Light:	Temperature:	Difficulty:
Central and South America	Moderate	70–77°F (21–25°C)	1 or 2

Specimen plants

These are unusual, either because of their appearance or because, in many cases, they are not rooted into the ground but securely attached to a support by means of thin nylon wire or cotton thread. These plants are very decorative and provide a strong contrast with less flamboyant foliage. However, some of them are quite difficult to grow.

Broadleaf anubias

This anubias can be planted in the fore-ground but it can also be attached with nylon wire to a rock or even a piece of wood (in which case the roots will eventually become anchored to the support). Whatever method is used, overly strong lighting must be avoided, so it cannot be placed near the surface. This plant thrives in slightly acidic, fairly soft water. It is a slow-growing species.

1 BROADLEAF ANUBIAS
Anubias barteri

Maximum size: **9 in. (24 cm)**

Origin:	Light:	Temperature:	Difficulty:
West Africa	Moderate to bright	72–82°F (22–28°C)	2

Java fern

This is one of the few ferns found in an aquarium. It is amphibious, but can be cultivated entirely underwater. It is relatively hardy once secured to a rock or piece of wood, but the roots of the rhizome must not be put into the ground because they are liable to rot. It does not like strong light, and appreciates fairly hard, alkaline water. The leaves appear to contain a substance that repels herbivorous fish. Over time, the leaves will begin to blacken; remove the affected areas. Java fern can be propagated by dividing the rhizome. The pieces will go on to grow and produce leaves.

2 JAVA FERN
Microsorium pteropus

Maximum size: **10 in. (25 cm)**

Origin:	Light:	Temperature:	Difficulty:
Southeast Asia, Java	Restricted	68–79°F (20–26°C)	1

African (Congo) fern

This fern appreciates water movement, so is ideally suited to a position near the filter outlet. It must be fixed to a support. It grows slowly but can be propagated in an aquarium by cutting off the runners.

3 AFRICAN (CONGO) FERN
Bolbitis heudelotii

Maximum size: 12 in. (30 cm), sometimes more

Origin:	Light:	Temperature:	Difficulty:
Africa	Moderate	72–82°F (22–28°C)	3

Java moss

This amphibious moss can grow very well under water. It can be planted in the substrate or attached to a support, such as wood or rocks. It is useful in an aquarium because it provides a surface for fish to lay their eggs on, as well as shelter for both eggs and fry. However, the fine leaves can become choked by suspended debris; the water must therefore be meticulously filtered. It can be propagated by dividing and replanting clumps around the aquarium.

Madagascar laceleaf

The structure of the leaves is highly distinctive, as they consist entirely of leaf veins, but it sometimes has the advantage of trapping unwanted filamentous algae. This plant requires fairly soft and well-filtered water. There are species from this genus that sport leaves with a more orthodox form – *Aponogeton crispus* is one of the best known, and also the easiest to grow. Some aquarists have managed to reproduce these plants using flowers that grow outside the water, but this is difficult to achieve.

4 JAVA MOSS
Vesicularia dubyana

Maximum size: creeping species, 2 in. (5 cm) plus

Origin:	Light:	Temperature:	Difficulty:
Southeast Asia, Malaysia, Java, India	Undemanding	68–86°F (20–30°C)	1

5 MADAGASCAR LACELEAF
Aponogeton madagascariensis (A. fenestralis)

Maximum size: up to 16 in. (40 cm)

Origin:	Light:	Temperature:	Difficulty:
Madagascar	Moderate	68–75°F (20–24°C)	3

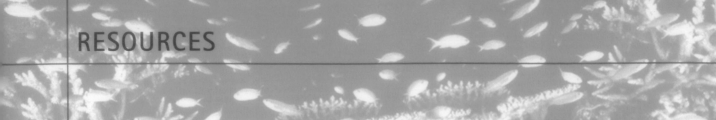

RESOURCES

PUBLIC AQUARIUMS IN THE U.S.

Aquarium of the Pacific
Long Beach, California
www.aquariumofpacific.org

Florida Aquarium
Tampa, Florida
www.flaquarium.org

Maui Ocean Center
Wailuku, Hawaii
www.mauioceancenter.com

National Aquarium in Baltimore
Baltimore, Maryland
www.aqua.org

National Aquarium in Washington
Washington, District of Columbia
www.nationalaquarium.com

New England Aquarium
Boston, Massachusetts
www.neaq.org

New York Aquarium
Brooklyn, New York
www.nyaquarium.com

Oregon Coast Aquarium
Newport, Oregon
www.aquarium.org

Seattle Aquarium
Seattle, Washington
www.seattleaquarium.org

Sonoran Sea Aquarium
Tucson, Arizona
www.tucsonaquarium.com

Steinhart Aquarium
San Francisco, California
www.calacademy.org/aquarium

Waikiki Aquarium
Honolulu, Hawaii
waquarium.otted.hawaii.edu

PUBLIC AQUARIUMS IN CANADA

Aquarium and Marine Centre
Shippagan, New Brunswick
www.gnb.ca/aquarium

Parc Aquarium du Québec
Quebec, Quebec
www.aquarium.qc.ca

Toronto Zoo
Toronto, Ontario
www.torontozoo.com

Vancouver Aquarium Marine
Science Centre
Vancouver, British Columbia
www.vanaqua.org/home

USEFUL ADDRESSES

Aquarium Hobbyist
*Online community for freshwater
and saltwater aquarists.*
www.aquariumhobbyist.com

Datafish
Aquarium fish database.
www.solodvds.com

FishBase
Database of fish species.
www.fishbase.org

Integrated Taxonomic Information System
*Database of plant and animal
species names.*
www.itis.usda.gov

Tropica Aquarium Plants
Database of aquarium plants.
www.tropica.com

ASSOCIATIONS

American Cichlid Association
www.cichlid.org

American Livebearers Association
www.livebearers.org

Aquatic Gardeners Association
www.aquatic-gardeners.org

Canadian Association of Aquarium Clubs
www.caoac.on.ca

Canadian Cichlid Association
www.cichlidae.ca

Canadian Goldfish Society
www.canadiangoldfish.com

Federation of American Aquarium Societies
www.gcca.net/faas

Goldfish Society of America
www.goldfishsociety.org

Marine Aquarium Council
www.aquariumcouncil.org

Marine Aquarium Societies
of North America
www.masna.org

North American Native Fishes Association
www.nanfa.org

Northeast Council of Aquarium Societies
www.northeastcouncil.org

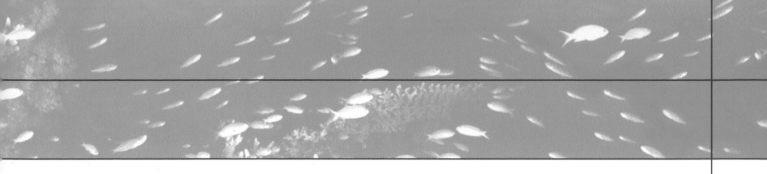

BOOKS

Andrews, Chris, Adrian Exell, and Neville Carrington. *Manual of Fish Health: Everything You Need to Know about Aquarium Fish, Their Environment and Disease Prevention.* Buffalo, NY: Firefly Books, 2003.

Axelrod, Herbert R., Warren E. Burgess, Neal Pronek, and Jerry G. Walls. *Dr. Axelrod's Atlas of Freshwater Aquarium Fishes*, 10th ed. Neptune, NJ: TFH Publications, 2004.

Bailey, Mary, Sean Evans, Nick Fletcher, Andy Green, Peter Hiscock, Pat Lambert, and Anna Robinson. *500 Ways to Be a Better Freshwater Fishkeeper.* Buffalo, NY: Firefly Books, 2005.

Burgess, Warren E., Herbert R. Axelrod, and Raymond E. Hunziker. *Dr. Burgess's Atlas of Marine Aquarium Fishes*, 3rd ed. Neptune, NJ: TFH Publications, 2000.

Dakin, Nick. *Complete Encyclopedia of the Saltwater Aquarium.* Foreword by Julian Sprung. Buffalo, NY: Firefly Books, 2003.

Dawes, John. *Complete Encyclopedia of the Freshwater Aquarium.* Buffalo, NY: Firefly Books, 2001.

Garratt, Dave, Tim Hayes, Tristan Lougher, and Dick Mills. *500 Ways to Be a Better Saltwater Fishkeeper.* Buffalo, NY: Firefly Books, 2005.

Goldstein, Robert J. *American Aquarium Fishes*, W. L. Moody, Jr. Natural History Series 28. With Rodney W. Harper and Richard Edwards. College Station, TX: Texas A&M University Press, 2000.

Kasselmann, Cristel. *Aquarium Plants.* Melbourne, FL: Krieger Publishing Company, 2003.

Rogers, Geoff, and Nick Fletcher. *Focus on Freshwater Aquarium Fish.* Buffalo, NY: Firefly Books, 2004.

Thraves, Stuart. *Setting Up a Tropical Aquarium Week-by-Week.* Buffalo, NY: Firefly Books, 2004.

MAGAZINES

Aquarium Fish Magazine
www.aquariumfish.com

Freshwater and Marine Aquarium Magazine
www.famamagazine.com

Practical Fishkeeping Magazine
www.practicalfishkeeping.co.uk

Tropical Fish Hobbyist
www.tfhmagazine.com

PICTURE CREDITS: